Concentrated Dispersions

ACS SYMPOSIUM SERIES **878**

Concentrated Dispersions

Theory, Experiment, and Applications

P. Somasundaran, Editor
Columbia University

B. Markovic, Editor
Columbia University

**Sponsored by the
ACS Division of Colloid and Surface Chemistry**

American Chemical Society, Washington, DC

Library of Congress Cataloging-in-Publication Data

Concentrated dispersions : theory, experiment, and applications / P. Somasundaran, editor, B. Markovic, editor ; sponsored by the ACS Division of Colloid and Surface Chemistry.

    p. cm.—(ACS symposium series ; 878)

    Includes bibliographical references and index.

    ISBN 0–8412–3834–0 (alk. paper)

    1. Colloids—Congresses

    I. Somasundaran, P., 1939- II. Markovic, B. (Berislav), 1957- III. American Chemical Society. Division of Colloid and Surface Chemistry. IV. American Chemical Society. Meeting (223rd : 2002 : Orlandl, Fla.) V. Series.

QD549.C65   2004
541′.345—dc22   2003062842

The paper used in this publication meets the minimum requirements of American National Standard for Information Sciences—Permanence of Paper for Printed Library Materials, ANSI Z39.48–1984.

PRINTED IN THE UNITED STATES OF AMERICA

Foreword

The ACS Symposium Series was first published in 1974 to provide a mechanism for publishing symposia quickly in book form. The purpose of the series is to publish timely, comprehensive books developed from ACS sponsored symposia based on current scientific research. Occasionally, books are developed from symposia sponsored by other organizations when the topic is of keen interest to the chemistry audience.

Before agreeing to publish a book, the proposed table of contents is reviewed for appropriate and comprehensive coverage and for interest to the audience. Some papers may be excluded to better focus the book; others may be added to provide comprehensiveness. When appropriate, overview or introductory chapters are added. Drafts of chapters are peer-reviewed prior to final acceptance or rejection, and manuscripts are prepared in camera-ready format.

As a rule, only original research papers and original review papers are included in the volumes. Verbatim reproductions of previously published papers are not accepted.

ACS Books Department

Contents

Indexes

Preface

The world of real colloidal dispersions in soils, emulsions, reproduction, and printing technologies, pharmaceuticals formulation, surface coatings, food and household products, and the like invariably involves concentrated or dense suspensions where the dispersed phase can be as large as 50% or more by volume. The classical experimental tools of light scattering or microelectrophoresis, for example, are of limited value in such systems and the classical Derjaguin–Lindau–Verwey–Overbeck (DLVO) theory of the stability of such systems ignores many body interactions.

Marked advances in experiment, theory, and technology have allowed the field of concentrated dispersions to progress significantly during recent years. A symposium *"Concentrated Colloidal Dispersions: Theory, Experiment, and Applications"* was conducted at the American Chemical Society meeting in Orlando, Florida in 2002 to review the current state of knowledge of this centrally important field in colloidal science and to explore new research directions.

Theorists, experimentalists, and technologists were brought together to review progress in the areas of dense dispersion systems; modeling and many body theoretical effects; basic and applied rheology; electrokineticeffects; and sensing and measurement of size, charge, and forces; structures in dense particulate systems; synthesis of special emulsion droplets and nanoparticles; confirmation of polymers; aggregation of particles; and dynamics of free and adsorbed polymer molecules in concentrated suspensions. Selected papers were reviews by experts in the field and were refined as required before the presentations.

We feel that this compilation presents a comprehensive examination of the progress made on concentrated dispersions along with problems and opportunities apparent through the chapters.

P. Somasundaran

NSF I/UCR Center for Surfactacts
Langmuir Center for Colloids and Interfaces
Columbia University
500 West 120th Street
New York, NY 10027
212–854–2926 (telephone)
212–854–8362 (fax)
ps24@columbia.edu (email)

Concentrated Dispersion

Chapter 1

Preparation and Characterization of Silver Nanoparticles at High Concentrations

Wei Wang and Baohua Gu

Environmental Sciences Division, Oak Ridge National Laboratory, Oak Ridge, TN 37831

Hydrophilic silver nanoparticles (5–30 nm) were prepared as hydrosols in the presence of a cationic surfactant, cetyltrimethylammonium bromide (CTAB), at relatively high Ag^+ concentrations (up to 0.1 M). The hydrophilic silver nanoparticles could be transferred to an organic phase by solvent exchange induced by inorganic salts, such as sodium chloride, with a high transfer efficiency (>95%). The hydrophobic silver nanoparticles are stable as concentrated organosols. They can be dried as powders for long-term storage and readily resuspended in a variety of organic solvents without loss of original particle integrity. Detailed characterization of these hydrosol and organosol particles was performed by transmission electron microscopy, dynamic lighting scattering (ζ potential), and UV-visible extinction spectroscopy. Infrared spectroscopic analysis provided evidence of the conformational changes of CTAB adsorbed on silver cores as the particles were transferred into organic solvents.

Introduction

Metal nanoparticles have been investigated intensively in recent years because of their size-dependent electronic and optical properties and the possibility of arranging them in micro- and nano-assemblies (1). In particular, a lot of effort has been devoted to the synthesis and characterization of stable dispersions of nanoparticles made of silver, gold, and other noble metals (2). Intriguing prospects for the development of novel electronic devices, electrooptical applications, and catalysis have been established (3).

Silver nanoparticles, mostly hydrosols, are perhaps most widely studied because of their important applications in catalysis (4) and photographic processes (5), and their roles in surface-enhanced Raman spectroscopy (SERS) (6). A rich variety of techniques are now available for producing silver nanoparticles as stable, colloidal dispersions in aqueous solution. These methods include reduction by ionizing radiation (7), photon- or ultrasound-induced reduction in solutions or reverse micelles (8), and chemical reduction in solution phase or in microemulsion (9), which is perhaps most widely used for silver colloid preparation. On the other hand, colloidal dispersions of silver in nonaqueous liquids— known to be difficult to prepare and to stabilize (10)— have received little attention. Previous studies also indicate that colloidal stability, particle size and morphology, and surface properties strongly depend on the specific method of preparation and the experimental conditions applied.

Wet-chemical synthesis methods usually produce stable silver colloids at Ag^+ concentrations below 0.01 M; above this concentration, silver colloids usually become unstable or form aggregates. Here, we report an approach for synthesizing silver hydrosols with a relatively high silver concentration (up to 0.1 M) in the presence of cetyltrimethylammonium bromide (CTAB). In addition, the silver nanoparticles in the hydrosol may be transferred to and concentrated in an organic solvent through an inorganic-salt-induced solvent exchange process. Nanaparticles dispersed in organic solvents could be dried and harvested by evaporating the solvent to obtain hydrophobic nanoparticles, which can readily be redispersed into organic solvents and still retain their colloidal integrity.

Experimental Section

1. Materials

$AgNO_3$ (≥99%) and NaCl (99%) were purchased from J. T. Baker. $NaBH_4$ (98%), cetyltrimethylammonium bromide (CTAB, ≥99%), chloroform (99.8%), and cyclohexane (99%) came from EM Science. Hexane (analytical grade) was obtained from Bio Lab. All chemicals were used as received. Water was of ~18 MΩ·cm resistivity, obtained from a Millipore Milli-Q plus water purifier.

2. Measurements

(a) Absorption spectra of the colloidal solution were taken with a HP 8453 spectrophotometer in quartz cell. For dried powder samples, a colloidal particle cast film was deposited on a quartz plate by evaporating the solvent.

(b) Direct imaging of the particles was obtained by a Hitachi HF-2000 electron microscope under an acceleration voltage of 200 kV. A drop of the silver sol was placed on a formvar/carbon film supported by copper grid (TED PELLA-LTD), and the solvent was allowed to evaporate. The average particle diameter <d> and its standard deviation σ were obtained by digitizing the printed micrographs and analyzing over 300 particles. The polydispersity is defined as the ratio σ/<d>.

(c) Infrared spectra were recorded at 4-cm^{-1} resolution with a Nicolet 760 Fourier transform infrared (FTIR) spectrometer equipped with a liquid nitrogen–cooled MCT detector. For pure CTAB, a portion of clear solution was deposited on the ZnSe window; after air-drying, the FTIR spectrum was recorded. For hydrophilic silver nanoparticles, the process was as follows: after centrifuging the colloid at 1200 rpm for an hour, the supernatant with excess CTAB was discarded and the collected wet solid nanoparticles were smeared on the ZnSe window. After air-drying, the spectrum of the nanoparticle cast thin film was taken. For hydrophobic nanoparticles, a drop of concentrated silver chloroform organosol was deposited on the ZnSe window, and the solvent was allowed to evaporate before the spectrum was collected. No smoothing was applied to the spectra.

(d) Zeta potentials of silver hydrosols were measured in a Brookhaven Zeta Plus light-scattering instrument.

3. Preparation of Silver Sols

(a) *Silver Hydrosols.* A typical preparation procedure involves the addition of one portion of $AgNO_3$ solution into another equal-volume portion of $NaBH_4$ solution containing CTAB (surfactant) at a molar ratio $[BH_4^-]/[Ag^+] = 2$ and $[Ag^+]/[CTAB] = 4$, with vigorous stirring. A dark brownish-gray colloid dispersion forms immediately upon the mixing, and with continuous stirring, the CTAB-stabilized colloid turns to a stable yellowish brown.

(b) *Silver Organosols.* Typically, 0.1 g solid NaCl is added to a mixture of 25 mL of the silver hydrosol stabilized by CTAB and 25 mL of an organic solvent such as chloroform, with vigorous stirring. A phase transfer is immediately induced, and the aqueous phase becomes colorless, while the organic phase is a colored colloidal dispersion.

Results and Discussion

1. Hydrosols Stabilized with CTAB

The reduction of $AgNO_3$ in solution with $NaBH_4$ is rapid, and the reaction can be written as

$$2AgNO_3 + 2NaBH_4 + 6H_2O = 2Ag + 2NaNO_2 + 2H_3BO_3 + 7H_2.$$

The reaction is usually performed in an ice-cooled water bath to reduce the reaction rate, and ionic/nonionic surfactants and polymers may be used as stabilizers in order to obtain stable silver colloids. Anionic surfactants have been used extensively in the preparation of silver colloids, but little attention has been paid to cationic surfactants (11). Pal et al. (11a, b) reported preparation of silver colloids in CTAB-based reverse micelles and aqueous solution; only large silver colloidal particles, ~ 65 nm in diameter, were obtained. Barnickel et al. (11c) prepared small silver nanoparticles in CTAB-based reverse microemulsion, but the size distribution of the silver particles was broad.

The stability and size distribution of silver colloids depend strongly on the relative concentrations of Ag^+ and reductant, the presence or absence of various stabilizers, and reaction temperature. In this study, we used a cationic surfactant, CTAB, as a stabilizer. We tried various preparation procedures and found that stable silver colloids formed at molar ratios of $[BH_4^-]/[Ag^+] = 2$ and $[Ag^+]/[CTAB] = 4$, with an initial Ag^+ concentration of $\sim 1 \times 10^{-1}$ to $\sim 1 \times 10^{-4}$ M. We also noticed that the order of chemical mixing affects the resulting colloid particle size and size distribution. The procedure that involved first mixing Ag^+ and CTAB and then adding the mixed Ag^+-CTAB solution to BH_4^- solution gave the most stable colloid dispersion and the narrowest particle size distribution. In this method, a burette was used to deliver the Ag^+-CTAB solution at a controlled, continuous flow rate directly into the BH_4^- solution with vigorous stirring. Simply pouring or adding the Ag^+ dropwise into the reaction mixture produced silver nanoparticles in a wide range of particle sizes. As stated previously, silver reduction reactions are usually performed at ice-cooled temperatures to reduce silver colloid aggregation because the formation and the growth of silver nanoparticles are sensitive to temperature. In the presence of CTAB, however, the reaction produced stable silver colloids even at room temperatures. Figure 1 shows representative transmission electron microscope (TEM) images of the silver nanoparticles produced at different initial Ag^+ concentrations and at room temperature. At low initial silver concentrations ($[Ag^+] = 5 \times 10^{-3}$ M), the colloids produce small sizes (<d> = 10.8 nm) and relatively narrow particle size distribution (polydispersity = 26%) (Figure 1a). With increasing initial silver concentration, the resulting particles become larger and are broadly distributed in size. For example, at $[Ag^+] = 2 \times 10^{-2}$ M, the average particle size and polydispersity were 12.3 nm and 35%, respectively (Figure 1b). However, the colloid dispersion is still stable and well-dispersed up to $[Ag^+] = 0.1$ M (Figure 1c), although the particles become bigger and the size distribution broadens (<d> = 22.2 nm, polydispersity = 53%).

5

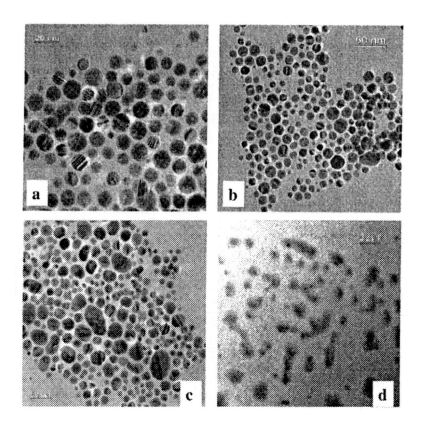

Figure 1. TEM images of CTAB capped Ag nanoparticles Synthesized at room temperature. (a) $[Ag^+] = 5\times10^{-3}$ M, $[CTAB]= 1\times10^{-3}$ M, $<d>=10.8$ nm, $\sigma= 2.9$. (b) $[Ag^+] = 2\times10^{-2}$ M, $[CTAB]= 4\times10^{-3}$ M $<d> = 12.3$ nm, $\sigma= 4.3$. (c) $[Ag^+] = 0.1$ M, $[CTAB]= 2\times10^{-2}$ M, $<d>=22.2$ nm, $\sigma= 11.8$. (d) $[Ag^+] = 2\times10^{-2}$ M, $[CTAB]= 0.1$ M, large particle lumps.

The dispersions of the metal nanoparticles usually display a very intense color due to plasmon resonance absorption, which can be attributed to the collective oscillation of conduction electrons that is induced by an electromagnetic field. In the UV-visible spectrum, a stable plasmon absorption band with symmetric shape was observed at 400 nm for the CTAB-stabilized silver colloids (Figure 2a). These spectra are the characteristic of well-dispersed

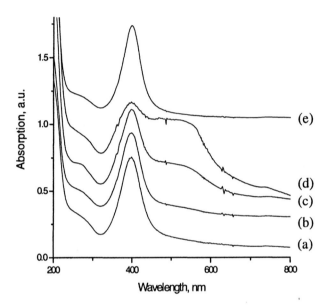

Figure 2. Normalized UV-visible spectra of CTAB capped Ag hydrosols prepared at initial concentrations of [Ag$^+$]=5×10^{-3} M and [CTAB]= 1×10^{-3} M, and diluted by water at (a) 25 times, (b) 50 times, (c) 75 times, (d) 100 times. Spectrum (e) was diluted 100 times by 1×10^{-4} M CTAB solution. All the spectra show absorption maxima at 400 nm.

silver colloid. As compared with the extinction adsorption of natively naked silver colloids at 390 nm (10e), the 10-nm red shift is due to absorption of CTAB on the silver nanoparticles. Zeta potential measurement shows a potential of +39 mV for the colloid, while the potential of native silver colloid is –46 mV. These results reveal that adsorbed CTAB cationic ion equivalents exceed negative charges on particle surfaces and thus have reversed the surface charge to positive. However, if the CTAB-stabilized colloid is diluted by water, the colloid color gradually changes from yellowish brown to bright yellow with no change in the absorption character of the spectrum (Figure 2b). With further dilution, the bright yellow color suddenly turns to pinkish yellow, and in the spectrum (Figure 2c,d), the absorption band exhibits a pronounced tail towards the longer wavelength, usually indicative of incipient particle aggregation (12). The corresponding Zeta potential decreases to ~+5 mV because CTAB molecules partially desorb into solution from the colloid particle surface. If the colloid is diluted by CTAB solution, no particle aggregation is observed in the spectrum (Figure 2e). During synthesis, we also noticed that the silver colloid is not stable; a gray precipitation of silver occurs when too low of a CTAB concentration is used. These results suggest that silver particles need enough CTAB molecules on the surface to keep their positively static stability. However, if the CTAB concentration is too high, the silver particles coalesce to form large lumps (Figure 1d), and the absorption band red-shifts to 520 nm.

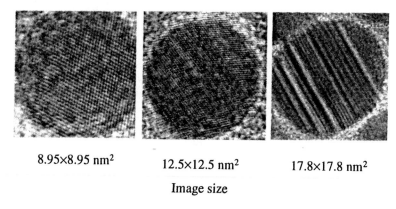

8.95×8.95 nm^2 12.5×12.5 nm^2 17.8×17.8 nm^2

Image size

Figure 3. High resolution TEM images of CTAB capped silver nanoparticles with different crystal structures: (a) single crystal, (b) multi-twin crystal, and (c) lamellar multi-twin crystal.

CTAB is an efficient stabilizer and strongly affects the silver particle sizes and size distributions. The silver colloids prepared in this work are stable and, in the absence of air, remain in aqueous dispersion without undergoing aggregation or flocculation for over 3 months, as indicated by the stable plasmon absorption band in UV-visible spectra. CTAB also exerts a dramatic influence on the structure of the silver particles. As seen in Figure 1, inhomogeneous imaging contrasts in the TEM pictures show imperfections in many silver particles. The silver nanoparticles synthesized in the presence of CTAB were a mixture of single crystalline particles (Figure 3a), multiple twinned particles (Figure 3b), and lamellar multiple twined particles (majority, Figure 3c). Heard et al. (13) and Henglein et al. (14) observed similar structures for anion capped silver colloids. High-resolution TEM images show that the imperfection of the particles arises from growth of crystallite in different directions. A possible explanation is that the cationic surfactant quickly forms a coating on negatively charged silver particles, and thus protects the particles from further growth and aggregation. Such an instantaneous sorption of CTAB on silver particles results in the formation of stable silver colloids at room temperature.

2. Organosols Stabilized with CTAB

One of the main challenges in the preparation and application of nanoparticles is how to transfer the particles into different physicochemical environments, such as fabricating nanoparticle-surfactant liquid crystal materials (15) and using them as catalysts for organic reactions in nonpolar solvents (16). Organosols of nanoparticles are attractive because they can be used to obtain packed structures of nanoparticles by self-assembly. Some solvent exchange methods have been developed to transfer silver hydrosols to organic solvents (10b, e, g). However, the transfer efficiency of anionic-surfactant-capped colloids into organic phase is usually low (<70%), and particle aggregation often occurs at the aqueous-organic interface. Here, we report that cationic-surfactant-capped silver colloids can be transferred quickly into an organic solvent, such as chloroform, with a high efficiency (>95%).

Experimentally, we first mixed silver hydrosol with an organic solvent and then added solid NaCl into the mixture. With vigorous stirring, the mixture becomes an emulsion, and the colloid phase transfer occurs immediately. The phase transfer rate depends on the organic solvents used and the quantity of salt added. When NaCl is added to the hydrosol in the proportion of 1 mg/mL, the phase transfer is completed in 1–2 minutes for chloroform and 3–5 minutes for cyclohexane. When the stirring is stopped, the emulsion reverts to two phases, a clear aqueous phase and a brown or yellow organic phase, the color depending on the concentration of the silver colloid. In addition to NaCl, many other cationic ions such as K^+, Mg^{2+}, and Ca^{2+} and anionic ions such as Br^-, PO_4^{3-}, and ClO_4^- may also serve as inducers for the phase transfer. During the phase transfer, the inorganic salt ion eliminates the CTAB outer layers in its bilayers and multilayers (see the spectral discussion in the next section). In the organic aqueous mixture, the desorbed and excess CTAB also may serve to emulsify nonaqueous liquid in colloidal aqueous dispersion, helping the CTAB-monolayer-coated silver nanoparticles transfer from water into the organic phase.

Use of small amounts of organic solvent (relative to hydrosol) can produce silver nanoparticles that are highly concentrated in the organic phases. These organosols could also be dried and harvested as powders by evaporating chloroform for storage purposes; the dried nanoparticles are readily redispersed into various organic solvents. As shown in Figure 4, absorption spectra of these organosols still retain symmetric and narrow absorption bands and are similar to those shown in Figure 2. However, the absorption maxima exhibit a red shift of 3–7 nm, depending on the specific organic solvents used for the colloid dispersion. An increase in the red shift with an increase in the solvent refractive index is expected from Mie theory (2a) and suggests that the silver particles are still well-dispersed in the organosols after the phase transfer and after drying and redispersion.

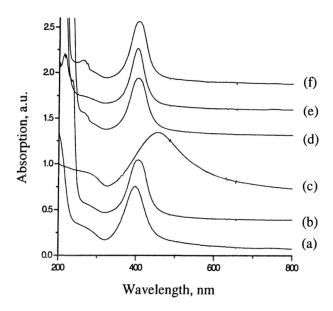

Figure 4. Normalized UV-visible spectra and peak positions of CTAB capped Ag colloids: (a) dispersed in water (400 nm), (b) in chloroform (406 nm), (c) dried particles from chloroform (455 nm), (d) dried particles redispersed in chloroform (406 nm), (e) redispersed in hexane (403) nm, and (f) redispersed in cyclohexane (406 nm).

3. Mechanistic Considerations of CTAB-Stabilized Silver Nanoparticles

FTIR spectroscopic analysis was performed to deduce the structure of adsorbed CTAB layers on silver nanoparticles and to understand the effects of CTAB on silver colloid stability. Figure 5 shows the FTIR spectra of pure CTAB and CTAB-capped silver nanoparticles in regions of 3100–2700 cm^{-1} and 1600–650 cm^{-1}.

In all the spectra in Figure 5, we can clearly observe the CH_2 and CH_3 stretching modes in 3000–2800 cm⁻¹, the CH_2 bending mode and the CH_3-(N^+) deformation mode in 1500–1450 cm⁻¹, and the CH_2 rocking mode in 735–710 cm⁻¹, which appear in spectra of aqueous and crystal CTAB (17). The spectra give unambiguous evidence of sorbed CTAB on both hydrophilic and hydrophobic silver nanoparticles.

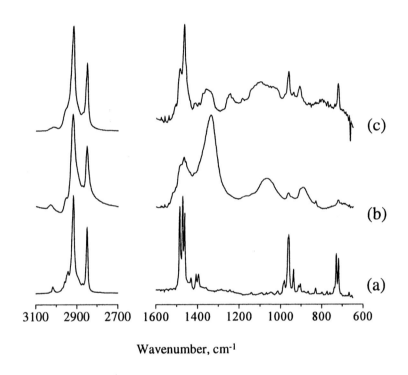

Wavenumber, cm⁻¹

Figure 5. FTIR spectra of (a) solid CTAB (b) CTAB capped hydrophilic Ag nanoparticles and (c) CTAB capped hydrophobic Ag nanoparticles.

The asymmetric and symmetric stretching modes of the head group CH_3-(N^+) are at 3017 and 2943 cm⁻¹, respectively, for solid CTAB (17b). The two bands markedly blue-shifted to 3130 cm⁻¹ and 2953 cm⁻¹ in the spectra of both hydrophilic and hydrophobic nanoparticles, suggesting that CTAB molecules were sorbed on silver nanoparticles via their $(CH_3)_3N^+$- head groups. The absorption peak at 960 cm⁻¹ in solid CTAB may be attributed to C-N stretching vibration. That the band shifts to 963 cm⁻¹ in the nanoparticle spectra supplies further evidence for the strong interaction between the $(CH_3)_3N^+$- head group and the silver surface.

The CH_2 stretching modes are very sensitive to hydrocarbon conformation and mobility (18). For pure CTAB, the CH_2 asymmetric and symmetric stretching modes appear at 2918 cm^{-1} and 2849 cm^{-1}, respectively. In comparison with pure CTAB, the two bands remain in the same location for hydrophobic nanoparticles, while they widen and shift to high frequencies at 2919 cm^{-1} and 2850 cm^{-1} for hydrophilic nanoparticles. These results imply that CTAB molecules have compact packing on hydrophobic nanoparticles. The blue shift of the two bands indicates that the CTAB molecular chains are more "fluid"—i.e., the molecular packing is not rigid on the hydrophilic nanoparticle surface. On the other hand, CTAB molecules may form a rigid monolayer on the hydrophobic particle surface, while additional adsorption on the monolayer results in the formation of loose bilayers or multilayers on the hydrophilic particle surface. In this case, the cationic ions have overbalanced the negative surface charges and formed stable, positively charged colloids, in coincidence with ζ potential measurement. Recently, El-Sayed et al. (19) also reported the formation of CTAB bilayers on gold nanorods synthesized by an electrochemical method.

The CH_2 scissoring and CH_2 bending bands are known to be very sensitive to intermolecular interaction and are therefore often used as key bands to check the state of packing of the methylene chain (20). For solid CTAB, the CH_2 scissoring mode gives a double structure at 1408 cm^{-1} and 1396 cm^{-1}, and the CH_2 bending mode gives another splitting doublet at 730 cm^{-1} and 719 cm^{-1}. The cause for band splitting is assigned to interchain vibrational coupling due to splitting of the factor group; band splitting accounts for a parallel packing of the trans-methylene chain in the orthorhombic unit cell . The two doublets are replaced by two single bands at 1466 cm^{-1} and 722 cm^{-1}, respectively, in CTAB-coated Ag nanoparticle spectra. These observations demonstrate that hydrocarbon chain interactions between molecules are weak and suggest that CTAB molecules adsorbed on silver nanoparticles may be arranged in monolayer or loose bilayer/multilayer forms.

Spectra of hydrophilic nanoparticles separated from hydrosols (by centrifuge) showed strong absorption bands around 1338 cm^{-1} and 1076 cm^{-1}. These absorption bands likely come from inorganic anionic ions, NO_3^-, NO_2^- and BO_3^{3-}. In order to remove all inorganic ions and excess CTAB in solution, we put the hydrosol into dialysis tubing against a large quantity of D.I. water. After two weeks of dialysis, all colloidal particles aggregated as black precipitates. In the infrared spectrum, all signals from inorganic ions also disappear, but narrowed CH_2 stretching bands still clearly remain. This suggests that some CTAB molecules might, as a monolayer, remain on the nanoparticle surface. The remaining CTAB monolayer on silver nanoparticles allows the hydrocarbon

12

chains to face the solution side and imparts a hydrophobic character to the particles. These results suggest an explanation for why the synthesis could not produce stable silver hydrosols at low CTAB concentrations. In other words, the surfactant molecules form a monolayer on the hydrophilic silver particle surfaces only at a low CTAB concentration. These surfactant molecules neutralize negative charges on the surface and thus reduce particle hydrophilicity and induce colloidal aggregates.

Conclusion

We have presented a convenient and easy-to-control procedure for preparing hydrophilic and hydrophobic silver nanoparticles with narrow size distributions in the range of 5 to 30 nm. These aqueous and organic colloid dispersions could be stored for at least 3 months without apparent colloidal aggregation or sedimentation. Dry powders of silver nanoparticles could be obtained from the organosols and readily redispersed into various organic solvents. The method is thus particularly suitable for large-scale industrial applications.

Acknowledgments

This research was partially supported by the Office of Basic Energy Sciences, U.S. Department of Energy, under contract DE-AC05-00OR22725 with Oak Ridge National Laboratory, which is managed by UT-Battelle LLC.

References

1.   (a) Ozin, G. A. *Adv. Mater.* **1992**, *4*, 612. (b) Alivisatos, A. P. *J. Phys. Chem.* **1996**, *100*, 13226; (c) Belloni, J. *Curr. Opin. Colloid Interface Sci.* **1996**, 2, 184; (d) Brus, L. *Curr. Opin. Colloid Interface Sci.* **1996**, 2, 197; (e) Fendler, J. H.; Meldrum, F. C. *Adv. Mater.* **1995**, 7, 607; (f) Schmid, G.; Chi, L. F. *Adv. Mater.* **1998**, *10*, 515; (g) Halperin, W. P. *Rev. Mod. Phys.* **1986**, *58*, 533.
2.   (a) Kreibig, U.; Vollmer, M. *Optical Properties of Metal Clusters*, Springer, Berlin, 1995; (b) El-Sayed, M. A. *Acc. Chem. Res.* **2001**, *34*, 257; (c) Henglein, A. *Chem. Rev.* **1989**, *89*, 1861; (d) Huang, M.; Choudrey, A.; Yang, P. *Chem. Commun.* **2000**, *12*, 1603.
3.   (a) Weller, H. *Angew. Chem., Int. Ed. Eng.* **1993**, *32*, 41; (b) Schmidt, G. *Chem. Rev.* **1992**, *92*, 1709; (c) Lewis, L. N. *Chem. Rev.* **1993**, *93*, 2693.

4. (a) Sun, T.; Seff, K. *Chem. Rev.* **1994**, *94*, 857; (b) Verykios, X. E.; Stein, F. P.; Coughlin, R. W. *Catal. Rev.-Sci. Eng.* **1980**, *22*, 197.

5. Mostafavi, M.; Marignier, J. L.; Amblard, J.; Belloni, J. *Radiat. Phys. Chem.* **1989**, *34*, 605.

6. Matejka, P.; Vlckova, B.; Vohlidal, J.; Pancoska, P.; Baumrunk, V. *J. Phys. Chem.* **1992**, *96*, 1361.

7. Henglein, A. *J. Phys. Chem.* **1993**, *97*, 5457.

8. (a) Henglein, A. *Chem. Mater.* **1998**, *10*, 444; (b) Ji, M.; Chen, X.; Wai, C. M.; Fulton, J. L. *J. Am. Chem. Soc.* **1999**, *121*, 2631; (c) Pethkar, S.; Aslam, M.; Mulla, I. S.; Ganeshan, P.; Vijayakrishnan, K. *J. Mater. Chem.* **2001**, *11*, 1710; (d) Salkar, R. A.; Jeevanandam, P.; Aruna, S. T.; Koltypin, Y.; Gedanken, A. *J. Mater. Chem.* **1999**, *9*, 1333; (e) Wang, W.; Asher, S. A. *J. Am. Chem. Soc.* **2001**, *123*, 12528.

9. (a) Creighton, J. A.; Blatchford, C. G.; Albrecht, M. G. *Trans. Faraday Soc.* **1979**, *75*, 790; (b) Fabrikanos, Von A.; Athanassiou, S.; Lieser, K. H. *Z. Naturforsch. B* **1963**, *18*, 612; (c) Frens, G.; Overbeek, J. Th. G. *Kolloid Z. Z. Polym.* **1969**, *233*, 922; (d) Henglein, A. *J. Phys. Chem.* **1979**, *83*, 2209; (e) Pileni, M. P.; Lisiecki, I.; Motte, L.; Petit, C.; Cizeron, J.; Moumen, N.; Lixon, P. *Prog. Colloid Polym. Sci.* **1993**, *93*, 1; (f) Toshima, N.; Yonezawa, T.; Kushihashi, K. *J. Chem. Soc., Faraday Trans.* **1993**, *89*, 2537; (g) Liz-Marzan, L. M.; Philipse, A. P. *J. Phys. Chem.* **1995**, *99*, 15120.

10. (a) Esumi, K.; Shiratori, M.; Ishizuka, H.; Tano, T.; Torigoe, K.; Meguro, K. *Langmuir* **1991**, *7*, 457. (b) Hirai, H.; Aizawa, H.; Shiozaki, H. *Chem. Lett.* **1992**, 1527. (c) Zeiri, L.; Efrima, S. *J. Phys. Chem.* **1992**, *96*, 5908; (d) Leff, D.V.; Brandt, L.; Heath, J. R. *Langmuir* **1996**, *12*, 4723. (e) Wang, W.; Efrima, S.; Regev, O. *Langmuir* **1998**, *14*, 602; (f) Wang, W.; Chen, X.; Efrima, S. *J. Phys. Chem. B* **1999**, *103*, 7238. (g) Huang, W.; Huang, Y. *Spectrosc. Spect. Anal.* **2000**, *20*, 449.

11. (a) Pal, T.; Sau, T. K.; Jana, N. R. *J. Colloid Interface Sci.* **1998**, *202*, 30; (b) Pal, T.; Sau, T. K.; Jana, N. R. *Langmuir* **1997**, *13*, 1481; (c) Barnickel, P.; Wokaun, A. *Mol. Phys.* **1990**, *69*, 1.

12. (a) Weitz, D. A.; Lin, M. Y.; Sandroff, C. *Surf. Sci.* **1985**, *158*, 147; (b) Quinten, M.; Schönauer, D.; Kreibig, U. *Z. Phys. D* **1989**, *12*, 521.

13. Heard, S. M.; Grieser, F.; Barraclough, C. G.; Sanders, J. *J. Colloid Interface Sci.* **1983**, 93, 545.

14. Henglein, A.; Giersig, M. *J. Phys. Chem. B* **1999**, *103*, 9533.

15. (a) Wang, W.; Efrima, S.; Regev, O. *J. Phys. Chem.* **1999**, *103*, 5613; (b) Chen, X.; Efrima, S.; Regev, O.; Wang, W.; Niu, L.; Sui, Z. M.; Zhu, B. L.; Yuan, X. B.; Yang, K. *Z. Sci. China B*, **2001**, *44*, 492.

16. (a) Andrews, M. P.; Ozin, G. A. *J. Phys. Chem.* **1986**, *90*, 2929; (b) Nakao, Y.; Kaeriyama, K. *J. Colloid Interface Sci.* **1989**, *131*, 186.

17. Wang, W.; Li, L.; Xi, S. (a) *J. Colloid Interface Sci.* **1993**, *155*, 369; (b) *Chin. J. Chem. Phys.* **1993**, *6*, 553.
18. (a) Casal, H. L.; Cameron, D. G.; Smith, I. C. P.; Mantsch, H. H. *Biochemistry* **1980**, *19*, 444; (b) Cameron, D. G.; Casal, H. L.; Mantsch, H. H. *Biochemistry* **1980**, *19*, 3665.
19. Nikoobakht, B.; El-Sayed, M. A. *Langmuir* **2001**, *17*, 6368.
20. (a) Synder, R. G. *J. Mol. Spectrosc.* **1961**, *7*, 116; (b) Umemura, J.; Cameron, D. J.; Mantsch, H. H. *Biochim. Biophys. Acta* **1980**, *602*, 32; (c) Wang W.; Li, L.; Xi, S. *J. Phys. Chem. Solids* **1993**, *54*, 73.

Chapter 2

Synthesis of Calcium Carbonate-Coated Emulsion Droplets for Drug Detoxification

Vishal M. Patel[1], Piyush Sheth[1], Allison Kurz[1], Michael Ossenbeck[1], Dinesh O. Shah[2], and Laurie B. Gower[1,*]

Departments of [1]Materials Science and Engineering and [2]Chemical Engineering, University of Florida, Gainesville, FL 32611

Nanoparticulate systems are being developed for use in pharmaceutical applications. Our goal is to synthesize "soft" emulsion particles coated with a porous "hard" inorganic shell which, when introduced to the blood intravenously, act as "nanosponges" for removing drug molecules from patients overdosed on lipophilic drugs. The synthetic approach utilizes a biologically inspired mineralization process of surface-induced deposition of calcium carbonate coatings onto charged emulsion droplets. Stearic acid and oil are dispersed in water as emulsion droplets, which are then coated with an amorphous mineral precursor to calcium carbonate using a polymer-induced liquid-precursor (PILP) process. Core-shell particles on the order of 1 to 5 μm in diameter have successfully been synthesized. Current experiments are directed at reducing the particle size using microemulsions, and templating porosity into the shell.

Introduction

The synthesis of micron- and nanoscale core-shell particulate systems, either hollow or fluid-filled, has been of considerable interest *(1)*. Core-shell particles find important applications in encapsulation of a variety of materials for catalysis and controlled release applications (e.g. drugs, enzymes, pesticides, dyes, etc.); for use as filler in lightweight composites, pigment, or coating materials; and in biomedical implant materials *(2-6)*. Recently, the use of particulate systems as a treatment for patients overdosed on lipophilic drugs has been proposed *(7)*. Several particulate systems, including microemulsions, polymer microgels, silica nanotubes and nanosponges, and silica core-shell particles, are currently being investigated for this detoxification purpose. It is proposed that, when intravenously administered to an overdosed patient, such particles will effectively detoxify the blood system of the particular lipophilic toxin by either:

- absorption, from the selective partitioning of the drug molecules from the blood to the hydrophobic core of the particle, or
- adsorption of the drug molecules onto lipophilic surfaces of surface functionalized particles.

Furthermore, in order to catalyze the toxin metabolism, and hence its removal from the blood, the immobilization of toxin-specific catabolic enzymes on or within particles will be pursued *(7)*. The work presented here focuses on the synthesis of oil-filled nanocapsules encapsulated by a porous calcium carbonate shell, which will remove the lipophilic drugs by absorption into the oily interior after passage through a porous mineral shell. The "hard" shell will provide stabilization to the emulsion, as well as impart some molecular screening, to avoid saturation of the particles with other lipophilic species in the blood. In light of the intended use, the design criteria include the following: the particles must be of nanoscale dimensions to avoid blockage of blood capillaries (and non-aggregating); the particles must be biocompatible (non-thrombogenic); and if the particles are not sufficiently small to pass through the blood-renal barrier, a biodegradable material is necessary for gradual removal of the particulates from the blood stream (at a pace slow enough for the body to tolerate the gradual release of the toxin). In addition, for this application it is ultimately desired to immobilize the environment-sensitive catabolic enzymes within the particles, in which case the synthesis must be accomplished under benign processing conditions.

Fabrication of hollow sphere particles has been accomplished by using various methods and materials. In general, three fabrication classes are currently employed: sacrificial cores, nozzle reactor systems, and emulsion or phase separation techniques *(1, 6)*. The first involves the coating of a core substrate with a material of interest, followed by the removal of the core by thermal or chemical means. In this manner, hollow particles of yttrium compounds *(8)*, TiO_2 and SnO_2 *(5)*, and silica *(6)* have been synthesized. Nozzle reactor systems make use of spray drying and pyrolysis, and their use has successfully led to the fabrication of hollow glass *(9)*, silica *(10)*, and TiO_2 *(11)* particles. Emulsion-mediated procedures are a third common method for hollow particle synthesis. This has been used to form latex *(2)*, polymeric *(12)*, and silica core-shell particles *(13)*.

Using the above techniques, a host of calcium carbonate coated core-shell particles have also been synthesized. By coating polystyrene beads with calcium carbonate, followed by removal of the polymer core, hollow particles in the 1 to 5 μm size range have been generated *(3, 14)*. Core-shell particles have also been synthesized using water-in-oil *(4, 15)*, and water-in-oil-in-water *(16, 17)* emulsions as templates for calcium carbonate nucleation. In other processes, Lee et al. *(18)* and Qi et al. *(19)* respectively use monolayer-protected gold particles and double-hydrophilic block copolymer (DHBC)-surfactant complex micelles as templates for calcium carbonate deposition, resulting in core-shell particles up to 5 μm in diameter.

We report herein a novel and facile method to synthesize calcium carbonate "hard" shell - "soft" core particles under benign conditions. Similar to those above, our process is a biomimetic one—in this case including an oil-in-water emulsion droplet as a template. The procedure relies on the surface-induced deposition of a calcium carbonate mineral precursor onto emulsion droplets by a Polymer-Induced Liquid-Precursor (PILP) process, elicited by including short-chained highly acidic polymers, such as polyaspartic acid, into crystallizing solutions of calcium carbonate which are slowly raised in supersaturation. The deposition of thin films of calcium carbonate onto glass coverslips using the PILP process has been demonstrated, as described previously *(20)*. In those studies, in situ observations revealed that the acidic polymer transforms the solution crystallization process into a precursor process by inducing liquid-liquid phase separation in the crystallizing solution. Droplets of a liquid-phase mineral precursor can deposit onto various substrates in the form of a film or coating, which upon solidification and crystallization, produces a continuous mineral film that maintains the morphology of the precursor phase (hence the name precursor). Thus, it was proposed that by using this PILP process to coat an oil droplet in solution, one could generate a fluid-filled core-shell particle with a thin uniform shell of calcium carbonate. In other studies, we have observed that this highly ionic PILP phase will preferentially deposit on charged regions of patterned substrates, which suggests that it might be possible to pattern porosity into the mineral shell by using an organic template with hydrophobic domains.

Unlike those particles reported previously, using the PILP process allows one to generate a smooth and uniform shell of calcium carbonate around the oil droplet, and not an aggregation of individual crystals, as is common among the work cited above. Furthermore, in this manner, oil can be encapsulated within the particle, leading to a "soft" fluidic core—a feature that is critical for the effective extraction of lipophilic molecules from aqueous media by an absorption mechanism.

Experimental Section

Emulsion Substrate Synthesis

Oil-in-water emulsion droplets were prepared by blending in a household kitchen blender, n-dodecane oil (Sigma-Aldrich) and distilled water in 1:9 volume ratio, stabilized with 1% w/v stearic acid (Sigma-Aldrich) (per oil phase volume). The distilled water was adjusted to the desired pH between 7 and 11 using 0.01M NaOH (Fisher Scientific) prior to emulsification.

Particle Synthesis

Immediately after preparing the emulsion, as indicated above, 1 mL of the emulsion was pipetted into 35 mm Falcon polystyrene petri dishes, followed by 1 mL of an 80 mM/400 mM $CaCl_2/MgCl_2$ solution (Sigma-Aldrich) (freshly prepared using distilled water, and filtered by 0.2 μm Acrodisc® syringe filters). Next, 36 μL of a freshly prepared and filtered 1 mg/mL solution of poly-(α,β)-D,L-aspartic acid (MW 8600) (ICN/Sigma-Aldrich) was transferred to each petri dish by micropipette. The petri dishes were then covered by parafilm, which was punched with a small hole, into which the outflow end of the tubing from an ultra-low flow peristaltic pump (Fisher Scientific) was inserted. At a rate of approximately 0.025 mL/min, 2 mL of a freshly prepared and filtered solution of 300 mM $(NH_4)_2CO_3$ (Sigma-Aldrich) was pumped into each petri dish (taking about 80 minutes to complete). The resulting product was collected and centrifuged at 8000 rpm for 10 minutes, rinsed with saturated $CaCO_3$ (Sigma-Aldrich), then recentrifuged under the same conditions. After a rinsing with ultrapure ethanol (Fisher Scientific), the product was recentrifuged a final time under the same conditions, and then left to dry in air overnight.

Particle Characterization

The dried particles were examined by an Olympus BX60 polarized light microscope, using a gypsum wave-plate in order to observe both amorphous and crystalline phases. For scanning electron microscopy (SEM) observations, particle samples were spread onto aluminum studs, and then gold-coated and examined with a JEOL 6400 SEM. Energy Dispersive Spectroscopy (EDS) was used for elemental composition analysis of the particle shell. For diffraction studies, dried particles were adhered to double-sided tape, and analyzed in a Philips APD 3720 X-ray instrument.

Results and Discussion

Several of the calcium carbonate core-shell systems discussed in the Introduction are generated by using a biomimetic process *(3, 4, 16, 17, 19)*. Mineralization in biological systems has been the focus of intense research because their successful mimicry has important implications for the synthetic design of superior materials. Exquisite control of mineral deposition in biosystems is thought to occur partly due to the presence of an insoluble organic matrix, along with modulation of the crystal growth process via soluble macromolecular species, such as acidic proteins and polysaccharides *(21)*. Thus, to mimic biological mineralization schemes, an organic substrate (i.e. an emulsion of some form, or a functionalized solid core) is used to direct the crystallization of a mineral shell in the fabrication of aforementioned calcium carbonated coated core-shell particles. In addition to the insoluble organic matrix, some groups include crystal inhibitors in the synthesis procedure. For example, the process used by Qi et al. *(19)* includes a portion of soluble DHBC molecules as inhibitors, which cooperate with the surfactant-copolymer micelle complexes as templates to allow the fabrication of hollow particles with a calcitic phase shell.

As a preliminary step to core-shell particle fabrication, and to better understand the deposition of calcium carbonate films on surfactant templates, we investigated the formation of freestanding films of the mineral under Langmuir monolayers spread at the air-liquid interface. Figure 1 shows polarized light micrographs of mineral films deposited under stearic acid monolayers. The micrographs were taken using a gypsum wave plate, which renders amorphous material to appear as the same magenta color as the background. As seen by the lack of birefringence in Figure 1A, the initial film is amorphous and optically isotropic (iso). Interestingly, the film cracked like a brittle glass when scooped onto a coverslip, which to our knowledge, is not typical for an amorphous calcium carbonate (ACC) phase (granular ACC precipitates are produced from

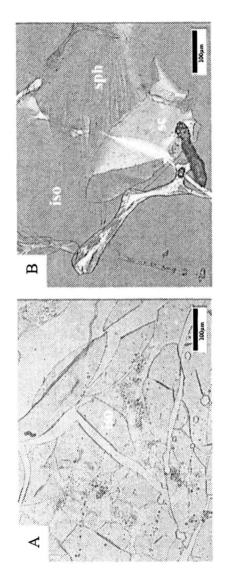

*Figure 1: Polarized light micrographs of thin calcium carbonate films
deposited under stearic acid monolayers via a PILP process. Bar = 100 μm.*

highly supersaturated solutions). If the films are removed from solution and let to dry in air, they crystallize in either spherulitic (sph) or single-crystalline (sc) patches (Figure 1B). Similar results were obtained under arachidic acid monolayers.

Repeating this experiment using cholesterol or diolein surfactants, in contrast, did not yield the uniform mineral film under the monolayer. Both stearic acid and arachidic acid surfactants have partially deprotonated carboxylic acid headgroup functionalities, while cholesterol and diolein surfactants, which bear alcohol moieties, remain polar but uncharged. Therefore, the surface charge on the monolayer is thought to play an important role in attracting mineral species and the ion-binding polymer to the surface, serving to increase ion saturation, and induce the deposition of the mineral precursor.

Considering these findings, the surface-induced deposition of a mineral shell onto a charged emulsion droplet was subsequently pursued. Using stearic acid as a surfactant, n-dodecane oil was dispersed in water to form an oil-in-water emulsion. To coat these emulsion droplets, they were first combined with Ca^{2+} dissolved in aqueous solution, along with polyaspartic acid to induce the PILP process. Mg^{2+} ions were also added to enhance the inhibitory action of the polymer, which helps to inhibit traditional crystal growth from solution (as opposed to from the precursor phase). The CO_3^{2-} counterion was subsequently pumped into the above mixture using ultra-low-flow peristaltic pumps. To monitor its effect on mineral deposition, the surface charge on the surfactant layer was varied by adjusting the pH of the aqueous solutions between 7 and 11 (pK_a of stearic acid is 10.15).

Figure 2A shows freshly coated particles synthesized in this manner at pH 7. As detected from the lack of birefringence under cross-polarized light, the particles, as expected, initially had an amorphous $CaCO_3$ shell. After rinsing the particles with saturated $CaCO_3$ and ethanol, the particles were allowed to dry in air. Figure 2B shows particles synthesized at pH 8 that were allowed to age in air for 1 week. The presence of birefringence in some of the spherical shells can now be detected, indicating an amorphous to crystalline phase transformation had taken place in the mineral shell, as was observed in the thin flat films. Furthermore, the Maltese cross pattern in the birefringence (see Figure 2B inset) indicates a spherulitic crystalline structure of the shell. The polycrystalline nature of spherulites evokes the possibility that the shell may be naturally porous (without requiring patterning of the deposition process), although it is at a very fine scale since the particles appear smooth at relatively high magnification. Dye-doped oil was used to determining the presence of oil in the core-shell particles, but the results were inconclusive due to the difficulty of examining particles of this size (which are small for optical microscopy, but too large for TEM).

Under scanning electron microscopy (SEM), the morphology and uniformity of the particles were better judged. From these observations, particles synthesized at pH 11 yielded the best results—fairly monodisperse, uniformly

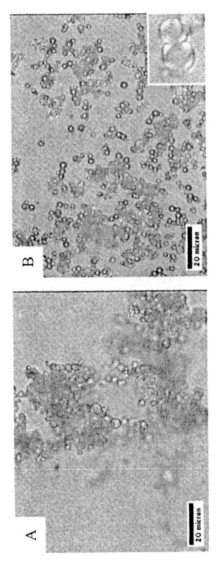

Figure 2: Cross-polarized light micrographs with gypsum wave plate of calcium carbonate coated emulsion droplets. Bar = 20 μm.

spherical particles of diameter ranging between 1 – 5 μm (see Figure 3A). Since the formation of a shell around the emulsion droplets was most enhanced at this pH setting, the increased surface charge at pH 11, compared to lower pHs, was deemed to be instrumental in the deposition of the PILP precursor. A sample of these particles was sheared between glass slides, and SEM of the resulting product is pictured in Figure 3B. The presence of spherical shell fragments and hollow cores (as indicated by the red arrow) confirms the core-shell structure of the fabricated particles. The shell thickness, based on SEM images such as those shown in Figure 3, was observed to be between 200 and 500 nm in thickness. No specific correlation between particle diameter and shell thickness was noticed, although it is thought that by controlling the reaction time, this property might be tailorable. The shell is a smooth uniform coating (Figure 3C), and although it appears to be spherulitic, it is not composed of an aggregation of individual crystals that nucleated from solution on the template, but rather it transformed from an amorphous precursor phase. Using Energy Dispersive Spectroscopy (Figure 3D), the presence of Ca, Mg, and O in the particle shell was confirmed, suggesting that the mineral is a Mg-bearing $CaCO_3$ phase (which is also seen for PILP films deposited onto solid substrates).

Current experiments are directed at detecting the presence of oil in the particle core, reducing the particle size using microemulsions, templating porosity into the shell, determining biodegradation properties, and *in vitro* testing of drug release (for drug delivery) or uptake (for toxicity reversal) capability.

Conclusions

Using highly acidic short-chained polymers to elicit a PILP process, thin shells of $CaCO_3$ have been deposited under a Langmuir monolayer at the air-water interface. To generate core-shell particles with an outer wall of $CaCO_3$, the same technique was employed to deposit the mineral precursor onto the surface of oil-in-water emulsion droplets, yielding "hard-soft" core-shell particles in the 1 to 5 μm diameter range. The smooth shells encapsulate an oil within the core, and are composed of a uniform homogenous mineral layer, and not an aggregation of crystallites. Our current efforts are directed at preparing nanoscale core-shell particles using microemulsion systems, and the templating of porosity into the mineral shells. Such particulate systems may find use in extraction processes such as uptake of overdosed drugs from the blood, or controlled-release applications.

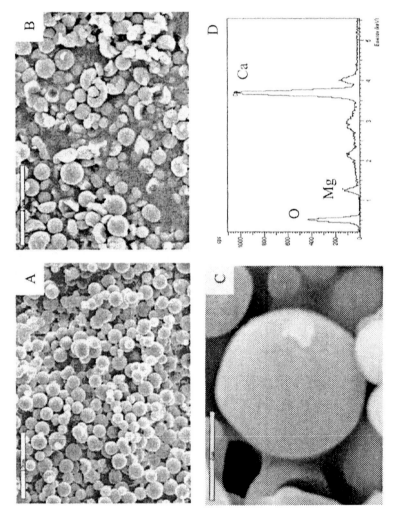

Figure 3: SEM (A,B, and C) and EDS (D) of calcium carbonate coated core-shell particles. Bar = 10 µm in A, B; and 2 µm in C.

Acknowledgements

We would like to thank Matthew Olszta and Sivakumar Munisamy for assistance with the particle characterization. This work was supported by funding from the Engineering Research Center (ERC) for Particle Science and Technology at the University of Florida under National Science Foundation Grant #EEC-94-02989, and its industrial partners.

Chapter 3

Colloidal Systems for Binary Mixtures Studies

P. Viravathana[1] and D. W. M. Marr[2]

[1]Department of Chemistry, Faculty of Science, Kasetsart University, Bangkok, 10900, Thailand
[2]Department of Chemical Engineering, Colorado School of Mines, 1500 Illinois Street, Golden, CO, 80401

Two novel colloidal systems are developed for experimental investigations of binary systems with short and long-ranged interactions. One, a "core-shell" system, is composed of a mixture of silica and silica-coated titania. In this system, the silica shell can be index-matched yet allowing the titania core to be manipulated by optical trapping. The other, relying on "contrast variation" system, is composed of an aluminosilicate synthesized with a slightly higher index of refraction than silica. By index-matching for silica, the aluminosilicate distribution can be determined. In the same way, by index-matching for aluminosilcate, the silica distribution can be extracted. The obtained results are compared to the results from thermodynamics calculations. With development of these model systems, there are new routes available for the optical trapping and light scattering based investigation of binary mixtures.

In the past few decades, there have been many studies on the phase behavior of binary systems, including mixtures of hard [1-5], long-ranged [6], and short-ranged [7-8] interactions via theoretical approaches; however, there have been relatively few experimental studies of these systems. In this work, we show that colloids provide a nice experimental system for the investigation of binary systems due to their wide range of sizes and the ease with which their interactions can be manipulated.

Wtih experiment, one needs to prepare colloidal systems that are appropriate for use with a specific technique. One technique we use is optical trapping, where one can directly manipulate individual colloids and for which there must be a mismatch in the index of refraction between the solvent and the colloidal particles. However, if we wish to minimize dispersion forces and study systems with short-ranged interactions, there must be an index-match between particles and solvent. Therefore, for binary studies using optical trapping techniques, two colloidal systems must be synthesized, one of which has a core with a different index of refraction than its shell. To pursue binary studies, interactions between different particles can be altered by varying solvent or manipulating surface charge. In doing so, optical trapping will still be feasible due to the mismatch in refractive index between the core and solvent.

In a "core-shell" system, when two different kinds of particles are dispersed in index-matching solvents, only the core having a different index of refraction will be seen via the optical microscope as illustrated in Figure 1. Once index matched, the core-shell particles can still be manipulated, moved around the suspension, or arranged in various patterns and sizes. By turning the beam on and off, the stability of such patterns can be determined and the associated phase behavior inferred. In addition, it is clear that such an approach could be used for the investigation of nucleation in binary mixtures.

The other technique we propose is light scattering. In order to perform binary studies using this approach, there must be two particle types having a slightly different index of refraction. If this difference is too great however, multiple scattering occurs, preventing determination of colloidal structure. After manipulating the interactions between particles, index matching for one species allows extraction of information on the behavior of the other. Since there must be some contrast in the system in order to use this approach, we call this "contrast variation".

In "contrast variation" for binary systems, once particles 1 are index-matched by the solvent, only particles 2 will scatter and information on particle 2 distribution can be obtained. By varying, the solvent mixture ratio, particles 2 can be index-matched and the particle 1 distribution can be determined. From both contributions, the intercorrelations between particle 1 and 2 can be extracted as shown in Figure 2.

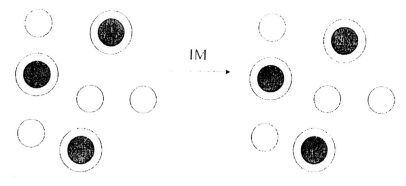

Figure 1 Core-shell approach IM = index matching.

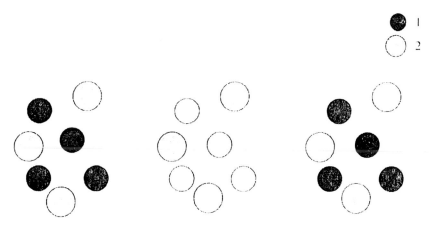

Figure 2 Contrast variation approach.

With development of these experimental systems, there will be new routes available for the investigation of binary mixture phase and nucleation behavior. Such studies will eventually lead to investigations with more complicated interactions and a higher number of components.

Core-Shell Approach

Since 1959, Iler patented the synthesis of core-shell particles [9], where there were discussions on particles with a silica shell and a core of different materials, including clays, talc, alumina, zinc sulfide, and titania. There have been several studies using a variety of techniques to study the chemical and physical properties of coated titanium dioxide surfaces [10,11] including the deposition of silica on the surface of titanium dioxide [12].

To directly manipulate colloidal systems, optical trapping techniques have been recently developed and are based on the use of radiation pressure to manipulate individual colloidal particles [13]. In this, a single focused laser beam is used to trap a small colloidal particle in 3-dimensions and to prevent it from settling against gravity or diffusing in its solvent. For stable trapping, the trap must overcome the scattering forces destabilizing the trap [14]. In addition to trapping a few particles, multiple particles can also be simultaneously trapped by using multiple laser beams or by using a single scanning laser beam as shown by Mio and Marr [15]. With this technique, one can directly manipulate ensembles of colloidal particles.

Experiments

Since the extent of the trapping force depends on the mismatch between the particles and solvent, index matching between colloid and solvent would minimize the van der Waals attraction but would not allow optical trapping. A core-shell system could make this technique usable if it were synthesized with materials of very different indices of refraction. A previous paper by Viravathana and Marr [16] describes the synthesis of silica-coated titania particles and estimates the required silica shell thickness needed to minimize the van der Waals attraction between spheres.

The methods used in the synthesis of silica have been discussed by Al-Naafa and Selim [17]. After obtaining both silica and silica-coated titania particles, the esterification was introduced by grafting polymer chains on the particle surface in order to prevent the aggregation.

Transmission electron microscopy was used to characterize the particle size. There were two sets of particle size of charged and esterified silica, one was approximately 144 nm in diameter and the other was approximately 357 nm in diameter. In this colloid synthesis, it was possible to obtain silica particles more monodisperse than titania particles. Because of the polydispersity of the titania cores however, estimation of silica shell thickness was difficult and was only determined approximately as 80 nm for a titania core diameter of approximately 565 nm.

Particle trapping was performed using a Nd:YAG laser of 532 nm at a power of 0.2 W. The beam was Gaussian in shape and vertically polarized with an initial diameter of 2.5 mm but focused down to a spot size of 1 μm for trapping. To create various optical trapping patterns, the beam was scanned using a piezoelectric mirror at a frequency of 300 Hz. Experimental details are presented by Mio and Marr [15] with sample preparation details found in Viravathana and Marr [16].

Results and Discussion

To synthesize monodisperse core-shell colloids we must lower the core polydispersity which we may be able to achieve by improving the quality of our reagents or through physical separation methods. In addition, we believe that addition of the silica shell may tend to lower the overall polydispersity as seen in the pure silica synthesis.

A picture of two core-shell colloids trapped and one core-shell colloid moving toward in order to complete a triangular pattern is shown in Figure 3. In order to verify that the silica shell was indeed index matched, mixtures of silica-coated titania and pure silica particles in the silica index-matching solvents were prepared. In such index-matching solvents the core-shell particles could clearly be seen; however, the esterified silica particles were only faintly visible in index-matched fluids and could not be trapped.

For binary systems, in mixed systems of silica and silica-coated titania particles in index-matching solvents, the trapping of the silica-coated titania was certainly possible but needed to be performed quickly before particles adhered to the glass slide due to its high settling velocity. Settling of the dense core-shell particles was perhaps the greatest experimental difficulty encountered in these experiments. Because a solvent of equivalent density is not available, future work should involve the synthesis of smaller colloids and the use of more viscous solvents to reduce settling velocities.

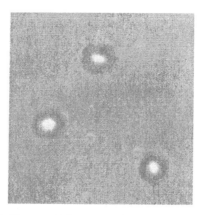

Figure 3 Core-shell titania/silica particles in shell index-matching solvent optically trapped and viewed under a microscope.

Contrast Variation

As discussed previously, to investigate colloidal systems using light scattering, there must be a difference in the index of refraction between colloids and their solvent. This is true for both single component and binary systems, where the index of refraction of both colloidal types must be different than the solvent index in order to scatter light. In addition, for a scattering approach to distinguish between two different kinds of colloids, there must be a difference in index of refraction between each. Our approach is to develop a system where one component is colloidal silica, chosen due to its availability in a large size range and the ability to prepare it with good uniformity. Since the other component in our mixture must have a slightly different index of refraction, we use a silica-based oxide due to its unique ability to readily form multicomponent silicates.

The syntheses of multicomponent silicates have been discussed by Brinker [18], where the preparation of SiO_2-Al_2O_3, SiO_2-B_2O_3, and SiO_2-TiO_2 has been briefly presented. The preparation of aluminosilicates has been of significant interest for the synthesis of mullite ($3Al_2O_3$-$2SiO_2$). Although there have been relatively few fundamental investigations on the hydrolysis and condensation of such materials, there has been some study of the synthesis of Al_2O_3-SiO_2 gels by hydrolysis of aluminosilicate ester at room temperature in acidic conditions [19]. Using these studies as a basis, we modify the preparation in order to form aluminosilicate sols, details of which are discussed in the following section.

Because of a different density and the presence of Al, aluminosilicate particles have a refractive index slightly higher than silica. By now dispersing mixtures of aluminosilicate and silica colloids in solvent mixtures of differing index of refraction, the system contrast can be carefully varied. When the solvent mixture has a refractive index close to silica, silica particles will not be visible and only aluminosilicate particles will scatter. When the solvent is at a mixture ratio now matched for aluminosilicate, only pure silica particles will scatter. This contrast variation method will allow the investigation of each component individually in a binary mixture.

Experiments

As discussed above, there have been studies involving the preparation of silica-alumina particles [19-21] including aluminosilicate gels [21]. By modifying the gel synthesis procedures, we have prepared aluminosilicate sols by the reaction of aluminosilicate ester, di-s-butoaluminoxytriethoxysilane, in water, catalyzed by ammonium hydroxide. Since the ester is immiscible in water, isopropanol must be present to act as a mutual solvent. The overall reaction occurs in two steps:

Step 1: Hydrolysis of the ester to acid:

$$\left(C_4H_7O\right)_2 Al - O - Si\left(OC_2H_5\right)_3 + H_2O \rightarrow xC_4H_7OH + yC_2H_5OH$$
$$+ \left(C_4H_7O\right)_{2-x}(HO)_x Al - O - Si\left(OC_2H_5\right)_{3-y}(OH)_y$$

x = 0,...,2 and y = 0,...,3

Step 2: Dehydration of the acid to form amorphous aluminosilicate:
$$\left(C_4H_7O\right)_{2-x}(HO)_x Al - O - Si\left(OC_2H_5\right)_{3-y}(OH)_y \rightarrow$$

$$\left(C_4H_7O\right)_{2-x} O_{\frac{x}{2}} Al - O - Si\left(OC_2H_5\right)_{3-y} O_{\frac{y}{2}} + \left(\frac{x+y}{2}\right)H_2O$$

x = 0,...,2 and y = 0,...,3

After obtaining both colloidal types, the esterification of silica and aluminosilicate particles was subsequently performed via the following surface reactions:

SiOH   +   ROH   ↔   SiOR   + H_2O
and
AlOH   +   ROH   ↔   AlOR   + H_2O

where, R = Octadecyl. To remove excess octadecanol, the suspension was typically centrifuged at 4,000 rpm with warm (40°C) cyclohexane to separate the esterified particles from the excess octadecanol, and the supernatant discarded. TEM and DLS were used to determine that esterified silica particles of approximately 360 nm diameter and aluminosilicate particles of approximately 240 nm diameter were obtained. The aluminosilicate was high polydispersity in particle size.

To study the contrast variation model, we use scattering techniques for which the basic equations describing x-ray, neutron, and light scattering are similar. The basic theory of small angle scattering is given by Guinier and Fournet [22] and Krause [23] and discussed by Viravathana and Marr [24]. Specifically, a wide angle light scattering (WALS) apparatus was constructed allowing measurement of the scattering intensity over a range of 60-120°, equivalent to a q range of 0.01-0.017 nm^{-1}.

To study binary systems, three different mixtures of known volume fraction of silica and aluminosilicate particles were suspended in solvents of different index of refraction. System 1 was of intermediate index solvent in which both particle types scattered light. System 2 is index-matched for silica (n = 1.45, density = 2.2) and system 3 is index-matched for the aluminosilicate silica (n = 1.49, density = 2.1±0.2).

In our studies, we used three solvents, including ethanol, benzene and cyclohexane and examined two kinds of particles, including charged and esterified colloids. Zeta potential measurements of silica and aluminosilicate particles in water show the decrease in surface potential after esterification. This illustrated the removal of some OH group at the particle surface and their replacement by octadecyl chains after the esterification reaction was performed. By removing surface charge through polymer grafting onto the surface, the charge-charge interaction between particles was significantly decreased and, when dispersed in nonpolar solvents, the particles hard-sphere like. For fully charged particles, OH groups at the surface induced a strong charge-charge interaction, providing long-ranged repulsions. When dispersed in nonpolar solvents, the charge-charge interaction decreased less than in the esterified systems.

Results and Discussion

In very dilute systems, the structure factor approaches unity and the scattering intensity is proportional only to the form factor. In order to measure the form factor and determine particle size, single component systems of dilute silica and aluminosilicate in ethanol were investigated. When the measured scattering curves were fitted to the form factor predicted for monodisperse spheres, particle sizes could be obtained. Similarly, the particle sizes of dilute mixtures of esterified silica and esterified aluminosilicate in various solvents could be measured and it was found that the particle size of the dilute mixture in benzene was higher than particle sizes from individual measurements. This indicated a tendency for particle aggregation as seen in previous studies [25] where it was mentioned that attractions were introduced in dispersions of hard spheres in benzene. For cyclohexane and ethanol solvents, particle sizes appeared smaller than in benzene, indicating the possibility of some aggregation.

Light scattering measurements at higher concentration allow the determination of structure factors. Theoretical prediction of hard sphere mixture correlation functions [26] can then be used to compare to our experimental measurements as shown in Figure 4 to 6. Figure 4 shows the measured structure

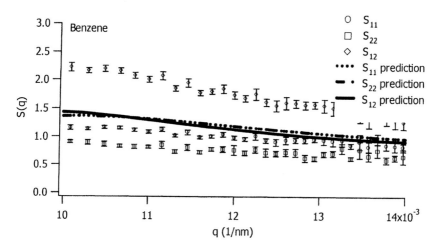

Figure 4 Binary mixture of esterified silica esterified aluminosilicate in benzene

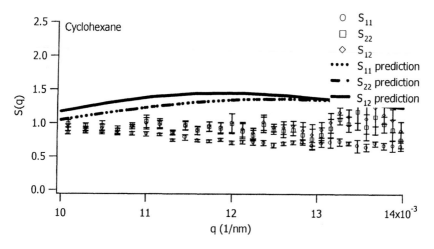

Figure 5 Binary mixture of esterified silica esterified aluminosilicate in cyclohexane

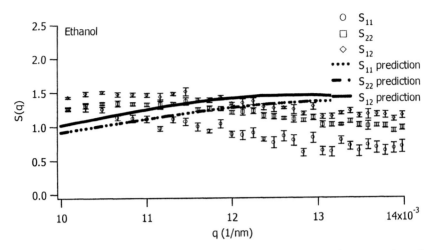

Figure 6 Binary mixture of esterified silica esterified aluminosilicate in ethanol

factors of a binary mixture of esterified silica and esterified aluminosilicate in benzene, where a steep slope in S_{12} is evident. This shows the difference in structure between system 2 (for S_{22}) and system 3 (for S_{11}). In system 1 (for S_{12}), the mixture is suspended in pure benzene where there is a possibility for aggregation to occur [25]. Whereas, in systems 2 and 3, ethanol is added to the mixture for index-matching, changing the solvent and slightly modifying the particle-particle interactions.

Our experimental results are lower in magnitude for S_{11} and S_{22} than results from thermodynamic calculation. This could be a result of our volume fraction values; in our experiments, the dispersion is initially dispersed but then sediments increasing the local volume fraction. As the laser passes through this location, the local volume fractions could be estimated. Predicted structure factors are computed using these estimated volume fractions leading to some error in the magnitude of S_{11} and S_{22}.

From Figure 5, the experimental structure factors and the results from calculation for the mixtures of esterified silica and esterified aluminosilicate in cyclohexane is achieved. This difference between the two could arise from the volume fraction approximation as mentioned earlier, or from fluctuations in experimental data at high q where the scattering intensity is low leading to large errors after subtracting the background.

In Figure 6, experimental results from the mixtures of esterified silica and esterified aluminosilicate in ethanol had similar magnitudes to predictions, however, there is a difference in the peak location. Here, we normalize the calculated curves by the radius obtained from the dilute mixed systems, which may not be accurate. If we adjust the radius, we can bring the results from both approaches into an agreement.

In the binary system with the long-ranged interaction, the mixtures of charged silica and charged aluminosilicate in ethanol, there were experimental difficulties due to large fluctuations from the measurements at low local density of the sample. OH groups at the silica surface will prevent close colloidal contact due to charge-charge interactions. Therefore, when light passes through the sample, it would result in less scattering. This hypothesis is based on the relatively large average lattice parameters determined for ordered single component silica colloids obtained from diffraction.

Conclusions

In this work we have chosen colloids for the development of experimental models for the investigation of binary systems. With their variable surface properties and dispersing behavior in solvents, colloidal particles can represent theoretical models with both short and long-ranged interactions. Two models have been developed to investigate the structure of binary systems using two different experimental techniques.

One technique we have used is optical trapping, which requires a mismatch in refractive index between particles and solvents. In order to have a short-ranged interaction however, the difference in index of refraction between particles and solvent must be minimized. Two different kinds of colloids were therefore prepared, including silica and silica-coated titania, known as a "core-shell" model. With the index of refraction mismatch between the solvents and the titania core, core-shell particles are trappable, and by index-matching the silica shell with the solvent, the interactions remained short-ranged. Preliminary investigations via optical trapping were performed in mixtures of silica and silica-coated titania particles suspended in silica index-matching solvents. It was found that core-shell particles could be clearly seen and were trappable; whereas, the silica particles were only slightly visible and could not optically manipulated.

In these studies, however, some difficulties were encountered. For example, due to the high density of the core-shell colloids, particles quickly settled and adhered to the glass surface. Since the particle density is so high, there is no equivalent-density solvent available. For future work, in order to overcome difficulties during trapping, smaller silica-coated titania particles and more viscous solvents should be used to decrease settling velocities.

The other technique we have investigated is light scattering, once again requiring an index-mismatch between particles and solvents. To experimentally investigate binary mixtures using this technique, one must have two colloidal types each with a slight difference in refractive index, known as a "contrast variation" model. For these studies we have selected silica and aluminosilicate.

Since their refractive indices are slightly different, by index matching for silica, information on the aluminosilicate distribution could be determined. In the same way, by index matching for aluminosilicate, the silica distribution could be studied.

To investigate binary mixtures, both kinds of colloids were dispersed in different solvents. Mixtures of esterified silica and esterified aluminosilicate were dispersed in benzene, cyclohexane, and ethanol. The behavior of esterified silica and esterified aluminosilcate mixtures in benzene was shown to be different than the same mixture dispersed in cyclohexane and ethanol. This was likely due to an aggregation that occurred in benzene, while there was no evidence for aggregation in cyclohexane and ethanol.

To validate some of our experimental observations of short-ranged interaction systems of silica-silica, aluminosilicate-aluminosilicate, and silica-aluminosilicate, we have compared our results to calculated values for binary hard sphere mixtures. It was found that mixtures in ethanol provided the best agreement, while mixtures in benzene and cyclohexane showed relatively large deviations from calculated values. For grafted-polymer stabilized colloidal particles, chain-solvent interactions are an interesting issue for further studies. These interactions can be altered by changing dispersing solvents or by changing temperature. In our studies, we have observed for these colloid types different behavior when they were dispersed in different solvents. Previous investigations have shown evidence of esterified silica in benzene showing an effective attraction on lowering temperature [25], changing particle behavior from hard sphere to adhesive hard sphere like. Such studies could be expanded for binary mixtures. By slowly changing the temperature of stabilized colloidal mixtures dispersed in a specific solvent, variation in the phase behavior as attractive interactions are slowly introduced could be studied.

Acknowledgments

P. Viravathana acknowledges the support from Department of Chemistry, Faculty of Science, and Kasetsart University on the grant for ACS conference.

References

1. Rovere, M., Pastore, G. *J. Phys.: Condens. Matter* **1994**, 6, A163-A166.
2. Melnyk, T. W.; Sawford, B. L. *Molec. Phys.* **1975**, 29(3), 891-902.
3. Trizac, E.; Eldridge, M. D.; Madden, P. A. *Molec. Phys* **1997**, 90(4), 675-678.
4. Lebowitz, J. L. *Phys Rev.* **1964**, 133(4A), A895-A899.
5. Denton, A. R.; Ashcroft, N. W. *Phys. Rev. A* **1990**, 42(12), 7312-7328.

6. Kerley, G. I. *J. Chem. Phys* **1989**, 91(2), 1204-1210.
7. Zaccarelli, E.; Foffi, G.; Tartaglia, P.; Sciortino, F.; Dawson, K. A. *Progr. Colloid. Polym. Sci.* **2000**, 115, 371-375.
8. Kolafa, J.; Nezbeda, I.; Pavlicek, J.; Smith, W. *Phys. Chem. Chem. Phys.* **1999**, 1, 4233-4240.
9. Iler, R. K. U.S. Patent 2,885,366, 1959.
10. Herrington, K. D.; Lui, Y. K. *J. Colloid Interface Sci* **1970**, 34, 447.
11. Furlong, D. N.; Rouquerol, F.; Rouquerol, J.; Sing K.S.W *J. Colloid Interface Sci.* **1980**, 75(1), 68.
12. Furlong, D. N.; Sing, K. S. W.; Parfitt, G. D. *J. Colloid Interface Sci.* **1979**, 69(3), 409.
13. Sato, S.; Inaba, H. *Optical and Quantum Electronics* **1996**, 28, 1-16.
14. Svoboda, K.; Block, S. M. *Optics Letters* **1994**, 19(13), 930.
15. Mio, C.; Gong, T.; Terray, A.; Marr, D.W.M. *Review of Scientific Instruments* **2000**, 71(5).
16. Viravathana, P.; Marr, D. W. M. *J. Colloid Interface Sci.* **2000**, 221, 301-307.
17. Al-Naafa, M. A.; Selim, M. S. *AIChE Journal* **1992**, 38(10), 1618-1630.
18. Brinker, C. F.; Scherer, G. W. *Sol-Gel Science*; Academic Press: San Diego, CA, 1990.
19. Pouxviel, P. C.; Boilot, J. P.; Dauger, A.; Huber, L. *Chemical Route to Aluminosilicate Gels, Glasses, and Ceramics*, in *Better Ceramics Through Chemistry II*; 1986.
20. Nishikawa, S.; Matijevic, E. *J. Colloid Interface Sci.* **1994**, 165, 141-147.
21. Boilot, J. P.; Pouxviel, J. C.; Dauger, A.; Wright, A. *Growth and Structure of Alumino-Silicate Polymers*, in *Better Ceramics Through Chemistry III*; 1988.
22. Guinier, A.; Fournet, G. *Small Angle Scattering of X-Rays*; Wiley, 1955.
23. Krause, R.; Arauz-Lara, J. L; Nagele, G. *Physica A* **1991**, 178, 241-279.
24. Viravathana, P.; Marr, D. W. M. Submitted for publication.
25. Rouw, P. W.; Woutersen, A. T. J. M.; Ackerson, B. J. *Physica A* **1989**, 156, 876-898.
26. Monson, P.A., Solution of the Percus-Yewick Equation for a Binary Mixture Using Gillan's Method, *PY-MIX*, **1984**.

Chapter 4

A New Wetting Polymer for Magnetic Dispersions: Methyl Acrylate and 2-Hydroxyl Ethyl Acrylate Copolymers

Meihua Piao[1], Shukendu Hait[2], David M. Nikles[3], and Alan M. Lane[1]

MINT Center and Departments of [1]Chemical Engineering and [3]Chemistry, The University of Alabama, Tuscaloosa, AL 35487
[2]Department of Chemistry and Geochemistry, Colorado School of Mines, 1500 Illinois Street, Golden, CO 80401

Abstract

A new wetting binder, methyl acrylate and 2-hydroxyl ethyl acrylate copolymer was synthesized and characterized for its ability to disperse magnetic particles in an organic solvent and compared to the commercial binder MR110. The effects of the amount of functional group and the binder molecular weight were studied. For this new binder, the functional group is a hydroxyl group that attaches to the particle surface. It is observed that the optimum amount of hydroxyl group is 5 mol%, and the molecular weight of the polymer has less significant effects on dispersion quality than the amount of functional group. The results suggest that the hydroxyl group is not such a strong functional group as the sulfonic group in MR110.

41

Introduction

The stability of a magnetic dispersion is determined by the adsorption of polymer molecules from solution onto particle surfaces and their conformations in the adsorbed state. The importance of the conformation of the adsorbed polymer was first recognised in 1951 by Jenckel and Rumbach, who suggested a " loops and trains" model to describe the adsorption behavior. It is also recognized that the segment density distribution is very important for steric stabilization. In other words, the distribution of trains, loops and tails is important for dispersion stability (1,2).

It is generally accepted that the interaction between polymer and particle surface is primarily a base and acid interaction (3-5). Therefore, in order to obtain a good dispersion, it is necessary to introduce functional groups in polymer molecules. A binder should have an optimum amount of functional groups. If the concentration of functional group is low, the train density is low resulting in low surface coverage. Consequently, this reduces the stability of magnetic dispersions. If the concentration of functional group is high, the functional group not only acts as an anchor group, but also exists in the loop and tail part of the adsorbed polymers, which causes gelation because of the strong hydrogen bonding interaction between the functional groups in the loops and tails (6).

According to the results of Kim et al. (7) and Sumiya et al. (8), a medium molecular weight polymer results in an improved and more stable dispersion. If the polymer molecular weight is low, the molecule is too short to provide good steric barrier since the loops and tails are reduced in this case. If the molecular weight is high, the long molecular chains cause entanglements or bridging effects hindering the orientation of particles, which is harmful to magnetic properties of the dispersion. Therefore, there exists an optimal molecular weight for binders.

MR110 is the most commonly used binder in commercial magnetic coatings. It has 0.7 wt% sulfonic acid groups, 0.6 wt% hydroxyl groups, and 3.0 wt% epoxy components. And the active functional group is sulfonic acid and hydroxyl which interact with the surfaces of the magnetic particles by acidic and basic interaction (9). From the point view of environment protection, MR110 is hazardous because it contains chloride. Our aim is to find an environmentally friendly polymer to replace the PVC copolymer as the binder in a magnetic dispersion system. Arcylate polymers are chosen because they do no harm to the environment and they dissolve in green solvents, such as ethyl lactate. In order to introduce functional groups, we used methyl arcylate and 2-hydroxyl ethyl acrylate (CPA), in which hydroxyl group is the anchor group. We regard MR110 ink sample as the model system to compare the quality of dispersions made by CPA polymers.

Experimental

2.1 Dispersion

The magnetic particles in this study were Co-γ-Fe$_2$O$_3$, with length 350 nm, density 4.8 g/cm^3 and aspect ratio 6. The binders were CPA polymers synthesized and characterized in our lab. The only difference between these CPA polymers is the monomer ratio, which leads to different concentrations of functional groups as shown in Table 1. Cyclohexanone was chosen as the solvent for its low volatility and its good solubility to CPA polymers. All the dispersions were milled in a ball mill for 48 hours.

There are two series of samples. The first series is used to investigate the optimum amount of functional group, including CPA1.5, CPA3, CPA5, CPA8, in which the numbers refer to the mole concentrations of functional groups. And the second series is for studying effects of molecular weight, including CPA5-01, CPA5-02, CPA5-03, CPA5-04, corresponding to molecular weight from low to high.

2.2 Characterization

Rheological measurements including linear viscoelasticity and steady shear viscosity were done with an ARES cotrolled shear strain rheometer with a truncated cone and plate geometry (50 mm diameter, 0.0392 rad cone angle). It was equipped with a solvent cover to minimize evaporation. The magnetic measurements were made using an AC susceptometer built in our laboratory.

Results

3.1 Optimize the concentration of hydroxyl group

Figure 1 shows the frequency dependence of storage modulus for the first series. All the samples are relatively independent of frequency. Storage modulus represents the elasticity of the system. The independence of G' on frequency means solid like behavior. For magnetic dispersions, this behavior results from network structure formed by the magnetic interaction between aggregates or flocs, which is counteracted by the steric barrier provided by polymer adsorption (10,11). The increase of polymer adsorption enhances the steric barrier that weakens the structure. If the polymer adsorption does not

Table 1. Functionality and inherent viscosity of CPA polymers.

Sample	Functionality (mol%)	inherent Viscosity (dl/g)
CPA1.5	1.5	0.49
CPA3	3.0	0.27
CPA8	8.0	0.39
CPA5 (CPA5-02)	5.1	0.57
CPA5-01	5.7	0.46
CPA5-03	5.1	0.66
CPA5-04	5.5	0.80
MR110		0.37

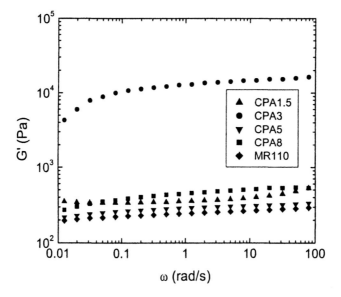

Figure 1. Frequency dependence of storage modulus of magnetic inks in the first series.

provide sufficient surface coverage, the particles form big aggregates. When the adsorption get worse, the dispersion become poorer or even not dispersed at all, leading to closely packed aggregates. In this situation, the G' value is expected to be extremely high due to its closeness to a solid. As one observed that MR110 ink has the smallest G' value resulted from relatively good polymer adsorption. CPA5 has the closest G' to MR110 ink. CPA1.5 has a higher G' than that of CPA5, which may come from the relatively big aggregates because of insufficiency of functional group. For CAP8, its excessive amount of functional group strengthens the network structure by strong hydrogen bonding between functional groups in loops and tails, which is responsible for its high G' value. The extremely high G' value of CPA3 could result from the closely packed aggregates, since it has both low functionality and molecular weight.

Steady shear viscosity data is represented in Fig. 2 and Fig. 3. It is observed that all CPA samples but CPA5 have shear thickening behaviors at relatively high shear rate. Especially, CPA8 has two shear thickening regions while others have one. MR110 shows typical shear thinning behavior. The shear thickening results from big aggregates in the system, which can be explained by the order to disorder transition mechanism (12). In the case of CPA8, the second shear thickening may be from the shear induced gelation due to its excessive amount of functional group. When the distance between aggregates becomes close, the opportunity for collision of the functional group in loop and tail increases, leading to gelation.

Transverse susceptibility characterizes the dispersion quality by probing the way the material responds to the external magnetic fields. Individual particles or small aggregates correspond to the change of field easily, which increases the magnitude of transverse susceptibility and enhances the height of peak. If the particle loading is similar, which means the impact from neighboring particles is similar, the high transverse susceptibility means more individual particles or small aggregates, which in turn means good dispersion. The details of this equipment can be found in (13). In Fig. 4, the transverse susceptibility result is illustrated, where one can see that the magnitude of transverse susceptibility of MR110 is the highest. Then come CPA5, CPA8, CPA1.5 and CPA3 respectively. It is reasonable to say that CPA5 has the best dispersion among CPA inks although it is worse than MR110 ink. One thing needs to be paid attention is that CPA3 has no peak, which is similar to the profile of dry magnetic powder. Thus, CPA3 has the worst dispersion quality, which is consist with the rheological results.

From the results above, we can conclude that CPA5 is the best binder among all these CPA polymers, although it is inferior to MR110. This result is consistent with the result obtained by Nakamae et al.(6). They found that at the optimum value of the magnetic properties, the amount of functional group was 4-6 mol% for OH group.

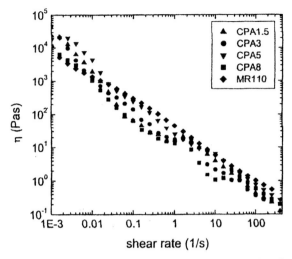

Figure 2. Steady shear viscosity of magnetic inks in the first series (plotted together).

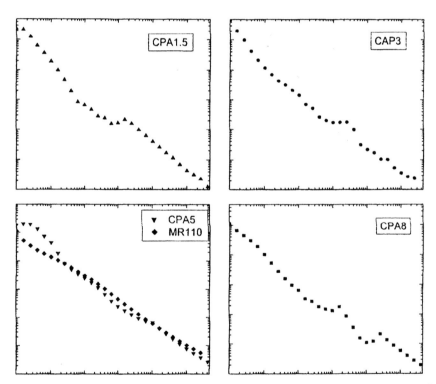

Figure 3. Steady shear viscosity of magnetic inks in the first series (plotted individually).

3.2 Effect of polymer molecular weight (MW)

When looking at the inherent viscosity data (Table 1), we find that inherent viscosity of CPA5 polymer is not only the highest among these CPA polymers, but also higher than that of MR110. It provides us a clue that it is possible to improve the properties of CPA5 polymer as a binder for magnetic dispersion by changing its molecular weight. To probe these results in detail, several CPA polymers with 5 mol% of OH group and different molecular weight represented by inherent viscosity were synthesized as shown in Table 1, where CPA5-01 has the lowest inherent viscosity, CPA5-04 has the highest, and those of CPA5-02 and CPA5-03 are in between.

In Fig. 5, the frequency dependence of storage modulus is depicted. The G' of all these ink samples are very similar both in magnitude and the independence of frequency, which means they may have network structure and the strength of their structures is similar. But their G' values are still larger than that of MR110 ink. Referring to the discussion in section 3.1, this implies the dispersion quality of MR110 is better than that of CPA polymers.

Figure 6 represents the steady shear viscosity of the second series. The magnitudes of the steady shear viscosity of all CPA5 inks are close. And CPA5-01, CPA5-03 and CPA5-04 show a shoulder at about the same shear rate. Big aggregates are hard to be orientated under shear. With the increment of shear rate, the structure perpendicular to shear direction begins to break. Since big aggregates encounter higher resistance, leading to the small shoulder in viscosity profiles. Compared to the first series, there is no shear thickening, which indicates that size of aggregates of the second series is smaller than that of the first series. The following shear thinning comes from the orientation and breakage of aggregates.

Figure 7 describes the transverse susceptibility results for the second series. CPA5-02 has the highest transverse susceptibility, then come CPA5-01, CPA5-03 and CPA5-04 respectively. As one can see from Table 1, CPA5-02 and CPA5-03 have same amount OH group while having different molecular weight. In their case, high MW results in low transverse susceptibility. And CPA5-01 and CPA5-04 have similar amount of OH group but different MW. Again, higher MW one has lower transverse susceptibility. Although CPA5-01 has lower MW, its OH group is more than that of CPA5-02, which results in its lower transverse susceptibility than that of CPA5-02.

Discussions

For the first series of experiments, the results indicate that 5 mol% is the optimal amount for OH group. If the amount of functional group is too

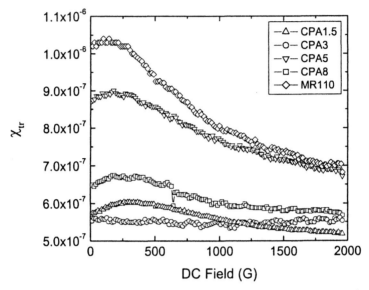

Figure 4. Transverse susceptibility of magnetic inks in the first series.

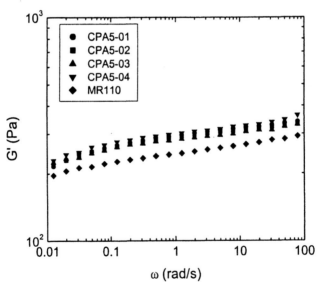

Figure 5. Frequency dependence of storage modulus of magnetic inks in the second series.

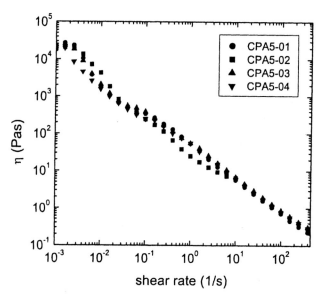

Figure 6. Steady shear viscosity of magnetic inks in the second series.

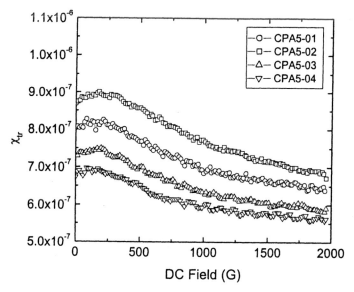

Figure 7. Transverse susceptibility of magnetic inks in the second series.

small, the train density decreases. Then the insufficient surface coverage of particles produces big aggregates in the system, which results in the shear thickening. While there exists excessive amount of functional groups, it causes gelation leading to high G' value and shear thickening at high shear rate as in the case of CPA8. But comparing to low functionality sample, the surface coverage of high functionality sample is better which results in smaller aggregates than the former, as represented by higher transverse susceptibility. In the case of CPA3, since both of its amount of functional group and molecular weight are low, the dispersibility of binder decrease a lot. Thus, the ink dispersed by CPA3 is similar to dry power depicted by transverse susceptibility test.   For the second series, functional groups of all these binders are close to the optimal amount, which makes them show similar behavior at rheological measurements. However, the difference is still observed from magnetic tests. When comparing CPA5-01 with CPA5-02, the amount of functional group has more significant effect on dispersion than MW, which is conformed by the relation of the amount of functional group with MW in the first series. When the amount of functional group is close, the high MW binder has low dispersibility. It may imply that decrease MW could improve dispersion. But there should be optimum MW because if MW is too low, it can not provide sufficient steric barrier. Our future work will include finding the optimum MW.

The polymer we synthesized is inferior to MR110 as a binder in the magnetic dispersions. There may be two reasons. One is the content of functional group. MR110 has various functional group i.e. SO_4, epoxy and OH, while CPA polymers have only OH group. OH is much weaker than SO_4 as a functional group, which means its interaction with particle surface is also weaker. And SO_4 occupy larger area than OH, which aids to increase the steric barriers. Sakai et al. (15) mentioned that when polymers are doped with strong functional group like sulfonic acid, the dispersion quality is improved. It is very possible that the inferiority comes from the weakness and smallness of functional group. The other reason is the higher MW of CPA5 than that of MR110. But according to the experiments, the effects from MW are not as significant as concentration of functional group. Thus, this is not the determinant factor.

Conclusions

From discussion above, it is conclude as follows:
(1) The concentration of functionality has more significant effect than molecular weight on dispersions.
(2) When the amount of functional group is same, low molecular weight binder has better dispersion quality than high molecular weight one.
(3) In order to obtain a good dispersion, the functionality and molecular weight should reach their optimal values.

(4) The weakness and smallness of hydroxyl group may be the main reason for the inferiority of CPA polymer to MR110 as a binder for magnetic dispersions. In order to have comparable dispersion quality with MR110, it may be necessary to introduce some strong functional group in CPA polymers.

Acknowledgement

The authors gratefully acknowledge financial support provided by the National Science Foundation through the Materials Research Science and Engineering Center at The University of Alabama (DMR-9809423) and the industrial sponsors of the Materials for Information Technology Center at The University of Alabama.

Reference

1. Tadros, Th. F.; *The Effect of Polymers on Dispersion Properties*; Academic Press, 1982.
2. Eirich , F. R.; *The Effect of Polymers on Dispersion Properties;* Academic Press, 1982.
3. Fowkes, F. M.; Huang, Y. C.; Shah, B. C.; Kulp, M. J.;T.B.lloyd, T. B.; *Colloids and surfaces* 1988, 29, 243.
4. Gooch, J. W.; *Polymers in Information Storage Technology*; Plenum Press: New York, 1989, 273.
5. Chen, W. J.; Wong, S. S.; Peng, W. G.; Wu, C. D.; *IEEE transactions on magnetics* 1991, Vol. 27, No. 6.
6. K. Nakamae, K.; Tanigwa, S.; Sumiya, K.; Matsumoto, T.; *Colloid & Polymer Science*, 1988, 266, 1014.
7. Kim, K. J.; Glasgow, P. D.; Kolycheck, E. G; *J. of Magn. Magn. Mater.* 1993, 120, 87.
8. Sumiya, K.; Taii, T.; Nakamae, K.; Matsumoto, T.; *Kobunshi Ronbunshu* (English edn.) 1981, 38, 123.
9. Jeon, K. S.; *Magnetic Particle Dispersion in Polymer Solution*, Dissertation, 1998.
10. Potanin, A. A.; Shrauti, S. M.; Arnold, D. W.; Lane, A. M.; *Rheol. Acta* 1998, 37, 89.
11. Kanai, H.; Navarrete, R. C.; Macosko, C. W.;L.E.Scriven, L. E.; *Rheol. Acta* 1992, 31, 333.
12. Bender, J.; Wagner, N. J.; *J. Rheol.* 1996, Vol. 40 No. 5, 899.
13. Shrauti, S. M.; *Magnetic and Rheological Characterization of Magnetic Inks*, Dissertation, 1999.
14. Sakai, H.; Cima, M. J.; Rhine, W. E.; *J. Am. Ceram. Soc.* 1993, Vol. 76, No. 12, 3136

Chapter 5

Techniques for Measurements in Concentrated Systems: Applications of Electromagnetic Scattering and Ultrasound

Vincent A. Hackley

Materials Science and Engineering Laboratory, National Institute of Standards and Technology, Gaithersburg, MD 20899–8520

Concentrated suspensions present a number of technical challenges for characterization and monitoring of the solid phase. Recent developments have provided the means to analyze many concentrated systems directly at solids loadings more closely resembling process conditions. This paper provides a general overview of four techniques: diffusing wave spectroscopy, ultra-small-angle x-ray scattering, acoustic spectroscopy and electroacoustics. Consideration is given to their particular advantages, limitations and applications.

The level of interest and amount of research activity in techniques for concentrated suspensions has grown rapidly during the past decade. Since many industrial applications of suspensions (see Table I) involve high solids loadings, there is a practical niche for measurement tools that can alleviate the need to dilute samples prior to analysis; dilution can shift chemical equilibria and may alter the physical state of the dispersed phase. Furthermore, there is a desire to have measurements that can be adapted to a process environment to permit real-

Table I. Some Key Industrial Applications of Concentrated Suspensions

Chemical mechanical planarization	Magneto-rheological fluids
Structural ceramic components for advanced engines	Physical sunscreen formulations
	Pigments and paints
Multilayer ceramic microelectronic packages	Inks and toners
	Phosphor coating slurries
Food emulsions & colloids	Pulp and paper products
Oil-water emulsions in petroleum industry	Cements
	Sol-gel applications

time monitoring of suspension properties. Given recent developments in both theory and instrumentation, these no longer represent unreasonable expectations.

In order to address these needs, techniques must contend with a number of challenging technical issues that can impact metrology in concentrated suspensions, such as particle interactions, multiple scattering effects, optical opacity, high viscosities, and a corrosive or abrasive nature. In addition, measurements are often required under dynamic conditions in order to monitor processes such as agglomeration, coalescence, and creaming.

A broad array of techniques is available for characterization of colloidal and fine particulate phases in highly loaded systems. A number of these methods are now commercially available, but many remain in the realm of academic research. Presently, it would not be possible to provide more than a cursory discussion of each method. Therefore, this paper will focus on just four techniques that should be of particular and broad interest: diffusing wave spectroscopy (DWS), ultra-small-angle x-ray scattering (USAXS), acoustic spectroscopy and electroacoustics. Except possibly USAXS, these techniques do not represent mature technologies; they are, in fact, still evolving to meet the demands of real world applications. Furthermore, each technique is capable of providing a different type of information relating to a different aspect of concentrated systems over various time and/or length scales.

This paper provides a brief overview of the underlying principles, uniqueness, limitations and required inputs for each technique. In each case, examples are given that demonstrate the unique capabilities.

Diffusing Wave Spectroscopy

Dynamic light scattering (DLS), or photon correlation spectroscopy (PCS), is a commercially established laser scattering method, widely utilized in the single-particle scattering regime to probe the dynamic properties of colloidal

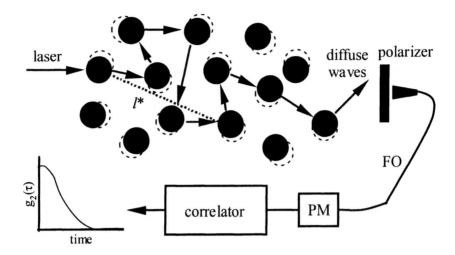

Figure 1. Stylized illustration of diffuse scattering and basic measurement set-up for DWS in transmission mode. Dotted circles indicate particle motion. FO: fiber optic, PM: photomultiplier. (after Ref. 8)

systems and to determine the hydrodynamic size of scatterers undergoing Brownian motion (*1*). While in conventional DLS the sample is highly dilute (and therefore nearly transparent), diffusing wave spectroscopy (DWS) extends DLS to more turbid media (*2, 3*). In fact, DWS exploits the multiple scattering condition of more concentrated systems by treating the transport of light through the medium as a diffusion process (see Figure 1). Using this approximation, the intensity autocorrelation function of diffuse scattering can be determined in backscatter or transmission geometry, and the dynamics of the underlying system extracted. DWS, a technique largely pioneered by the Weitz research group at Harvard, is appropriate for many concentrated weakly absorbing systems, including suspensions and polymer gels.

In DWS the diffusely scattered light is measured over short time intervals, of the order of ns to μs. The measured intensity autocorrelation function, $g_2(\tau) = \langle I(t)I(t+\tau)\rangle/\langle I(t)\rangle^2$, the Fourier transform of the Doppler power spectrum, can be related to the mean square displacement $\langle \Delta r^2 \rangle$ of the scatterers as a function of delay time τ. The characteristic hydrodynamic relaxation time τ_o is determined from analysis of the correlation function. For non-interacting colloidal systems, τ_o is related to the translational diffusion coefficient through the Stokes-Einstein equation, $D_0 = kT/6\pi\eta R_H = (1/\tau_o)k_0^2$, where η is the medium viscosity, R_H is the hydrodynamic radius and $k_0 = 2\pi n/\lambda$ is the wavenumber (*n* is the medium refractive index and λ is the laser wavelength in vacuum).

This relation holds for hard spheres (e.g., well screened electrostatic interactions) up to moderate concentrations (volume fractions of 10-20 %) (3). In more concentrated or strongly interacting systems, or for very small particles (where $k_0 \cdot d \sim 1$), interparticle hydrodynamic interactions and structural correlations can complicate the interpretation of DWS data. To account for these effects, hydrodynamic and static structure factors must be determined or calculated independently from the DWS experiment (4).

Since each photon is scattered by a large number of particles before reaching the detector, each particle must only move a very small fraction of a wavelength in order for a measurable intensity fluctuation to result. As a consequence, DWS probes particle motion on very short length scales of order 1 nm or less (5). Also, by virtue of the diffusion process, all angular dependence is lost, so DWS cannot provide specific information on the Q-dependence [scattering vector $Q=2\pi/\lambda$ (sin $\theta/2$), where 2θ is the scattering angle] of the dynamics. An important parameter in DWS is the transport mean free path l^*, which characterizes the average distance a photon must travel in the medium to randomize its direction (2). The so-called *scattering strength* $1/l^*$ depends both on particle size and volume fraction, and determines transmittance through the medium. The dependence of $1/l^*$ on size is weak for particles larger than about 100 nm (calculated for $\lambda \sim 500$ nm), but $1/l^*$ decreases rapidly below 100 nm, establishing a lower size detection limit for DWS (3).

DWS Applications

Two industrially important and well-studied applications for DWS are sol-gel transitions in colloidal suspensions and optical microrheology. Although commercial application as a research tool appears promising, DWS seems less practical for process applications.

Sol-Gel Transition

In colloidal gels, the scattering particles are localized to fixed averaged positions, but are still able to perform limited (sub-diffusive) motion over very small length scales (on the order of a few Å). In this case, the particles are nonergodic; that is, the measured time-averaged correlation function is no longer equivalent to the calculated ensemble-averaged correlation function (6), thus complicating interpretation. In order to achieve suitable ensemble averaging in nonergodic media, Romer and coworkers (6-8) at the University of Fribourg have recently developed a novel cell design.

Two slab-like glass cells are placed back-to-back with DWS in transmission mode. The first cell contains the nonergodic sample (i.e., gelling system under study), while the second cell contains an ergodic medium with slow exponential dynamics and moderate turbidity (7). The purpose of the second cell is merely to average the signal from the first cell. In their initial study, Romer et al. (7) used a 1.5% suspension of 810 nm polystyrene spheres in a water/glycerol mixture for the second cell. In the 2-cell configuration, the measured correlation function is the product of the individual correlation functions for the two cells. Therefore, the dynamics of the nonergodic system can be extracted by dividing the measured 2-cell autocorrelation function by the previously determined correlation function for the ergodic cell. Using this method, Romer et al. (7) followed the changing dynamics during enzyme-catalyzed gelation of a suspension of 298 nm polystyrene at a volume fraction of 20 %, from freely diffusing non-reactive particles to space-filling elastic gel. In a subsequent study (8), 40 % volume fraction alumina suspensions were destabilized by enzyme-catalyzed hydrolysis of urea. In this case, two modes of destabilization were examined: pH-shift and electrostatic screening, both of which can be induced by enzyme catalysis under appropriate conditions. For pH-shift, formation of a space filling gel network is indicated by the appearance of a plateau in the measured DWS autocorrelation function at long times. The plateau results from persistence of diffusional correlations due to the restrained motion of particles attached to the gel network. For gels formed by electrostatic screening, the plateau appeared prior to the percolation threshold. The authors attributed differences in dynamic behavior to variations in the aggregation kinetics associated with the two modes at different stages of the destabilization process.

Optical Microrheology

DWS can also be used to measure the microscopic viscoelastic behavior in complex fluids and elastic gels (5, 9-11). In one adaptation of this technique, tracer particles are mixed into an otherwise transparent polymer gel at concentrations that provide for diffusive scattering. The correlation function gives $\langle \Delta r^2 \rangle$ for the embedded particles as a function of the correlation time, from which the local storage $G'(\omega)$ and loss $G''(\omega)$ moduli can be calculated.

In a purely viscous medium, the tracer particles undergo free diffusion and $\langle \Delta r^2 \rangle$ should scale linearly with correlation time. Here, the local viscosity of the medium can be determined using the standard Stokes-Einstein relationship. By contrast, in a purely elastic (solid-like) medium, the particle motion will be constrained, and $\langle \Delta r^2 \rangle$ will reach a plateau value determined by the elastic modulus. In this latter case, the storage or elastic modulus is related inversely to $\langle \Delta r^2 \rangle$ and particle size of the tracer: $G'(\omega) \sim kT/\langle \Delta r^2_{plateau} \rangle$. In a visco-elastic medium, such as PVA gel, the material exhibits aspects of both types of behavior. In this case, frequency-dependent moduli can be obtained from the measured $\langle \Delta r^2 \rangle$ using a form of the generalized Stokes-Einstein equation (9). By combining DWS with DLS (using dilute tracers) the Weitz group has been able to obtain shear moduli covering 6 decades in frequency, and demonstrated excellent overlap with data obtained by macroscopic shear rheology over the limited experimental frequency range accessible by the latter technique (10).

Ultra-Small-Angle X-ray Scattering

Because of the penetrating nature of x-rays in condensed matter, measurements can be performed at relatively high solids concentrations. Like all diffraction or scattering phenomena, the angle of scattering varies inversely with the size of the inhomogeneity or feature giving rise to the scattering event. In conventional small-angle x-ray scattering (SAXS) the angles involved (less than a few tenths of a degree) permit analysis of structural features on a scale relevant to colloidal systems (roughly 1 nm to 100 nm). Coupled with appropriate functions for the particle form factor P(Q) and interparticle structure factor S(Q), particle size, size distribution and interparticle structure and periodicity can all be extracted from a single measurement or an appropriate combination of measurements (12). The development of maximum entropy methods has further improved the interpretation of SAXS data by reducing model bias and removing assumptions about the form of the particle size distribution (13).

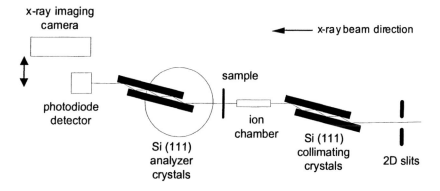

Figure 2. Double-crystal slit-smeared Bonse-Hart USAX camera at APS.

More recently, the development of ultra-small-angle x-ray scattering (USAXS) has extended the accessible length scales into the micrometer range, permitting the characterization of more complex systems with multiple levels of structure and the detection and characterization of oversize particles and aggregates (*14-16*). Although USAXS is available to researchers through large user facilities, it is not easily accessible for the routine analysis of materials. Irregardless, USAXS provides a unique research tool for investigative studies of concentrated systems and processes covering a range of length scales not easily achieved by other methods.

The National Institute of Standards and Technology USAXS facility at the Advanced Photon Source (APS) at Argonne National Laboratory, part of a co-managed user facility known as UNICAT (*15, 16*), consists of a double-crystal Bonse-Hart camera in slit-smeared configuration and an effective pin-hole collimated system for measurements in anisotropic systems (but with reduced intensity and range). The beam source is the 33ID injection device at the APL synchrotron. The distance from the source optics to the USAXS camera (shown in Figure 2) is about 60 m. Once in the camera optics, the beam passes through a pair of 6-reflection Si(111) collimating crystals, an ion chamber for monitoring beam intensity, and finally the test sample. The analyzer crystals are scanned through and away from the Bragg diffraction angle to give the scattered intensity as a function of the scattering vector Q. As a result of the multiple reflection geometry, the rocking curve width is less than 4 arc seconds, giving a band-pass of less than 1 eV, thus dramatically reducing the background within the ultra-small-angle scattering regime. The photodiode detector has such a wide range of linearity (many decades) that the direct beam intensity can be measured with and without the sample present, allowing for an absolute intensity calibration from first principles negating the need for scattering standards (*17*).

For the 33-ID injection device beam line, a continuous energy range from 4 keV to 40keV is available. However, the USAXS camera can accommodate from 5 keV to around 20 keV, and studies typically run at 10 keV corresponding to a wavelength of about 0.124 nm. The practical scattering vector range in the slit-smeared configuration is from about 0.0015 nm^{-1} to 5 nm^{-1}. This Q-range represents over 3 decades in length scale, from about 1 nm up to about 2 μm. In principle, higher Q values (smaller length scales) are possible with this setup, however in practice signal-to-noise issues reduce the upper range. The effective pin-hole setup, which uses an additional pair of collimating crystals, permits measurements in anisotropic materials and significantly simplifies the beam geometry and removes the need for desmearing, but at the cost of roughly a decade loss in both beam intensity and lower length scale cutoff (Q ~ 0.5 nm^{-1}).

The main advantages of using the synchrotron as an x-ray source, compared to a conventional rotating Cu anode, are two-fold. Much higher beam intensities are possible (around 4 x 10^{10} photons per mm^2 per second at the sample position), increasing sensitivity especially at the higher Q-range where the scattered intensity is generally low, and wavelength selectivity is possible.

USAX Applications

Computational methods to deal with particle size distribution and concentration effects, and also to recognize and interpret features associated with "characteristic" agglomerate and gel morphologies, continue to be modified and improved at various research institutions. Recently, these methods were applied to study the formation of agglomerates during the drying of yttria-stabilized zirconia nanosuspensions under varying pH conditions (18). The analysis of different Q-ranges in the slit-desmeared USAXS curves provides insight into both the agglomerate structure and the underlying formation mechanisms in this system. Scattering from primary particles, particle surfaces, agglomerate mass fractals and agglomerate surface structures can be identified in a single USAXS curve, providing structural information on many length scales. Similarly, pore size distributions, ranging from the nanopore regime to the macropore regime, have been characterized simultaneously in colloidal silica gels using USAXS (15). Inputs for USAXS are generally minimal, and may include the x-ray absorption cross-section for the test materials.

Acoustic Spectroscopy

Acoustic spectroscopy, in the present context, refers to the measurement of the attenuation and/or velocity of compression waves in multiphase fluids at ultrasonic frequencies. These frequency-dependent parameters are characteristic of the propagating material and can be related to its physical properties (e.g., particle size, density, composition, viscosity). There are certain analogies with optical absorption and scattering; however, one key difference is that sound waves interact much more weakly with condensed matter, thus providing penetrating capabilities for analyzing concentrated opaque suspensions.

The interactions of high-frequency sound with suspensions containing a dispersed solid or semi-solid phase can be very complex in nature, and may involve multiple loss mechanisms. Dukhin and Goetz (19) have summarized the possible loss mechanisms as follows: 1. *visco-inertial*: dissipation via viscous drag and inertia due to density contrast between dispersed and continuous phase; 2. *thermal*: thermodynamic coupling of pressure and temperature leads to dissipation as heat when differences in thermal properties exist between dispersed and continuous phase; 3. *scattering*: redirection of acoustic waves away from receiver; 4. *intrinsic (bulk)*: loss due to molecular level interactions in the pure phase; 5. *structural*: dissipation attributed to network oscillations, for example in gels; 6. *electroacoustic*: coupling of the electric and acoustic fields in suspensions of charged particles.

Fortunately, in many cases, a single loss mechanism dominates attenuation to the extent that the others can be ignored. Electrokinetic losses are always insignificant in this respect. Structural losses, although ill-defined, are probably only important in highly interconnected structures and at dense particle packings, and are thus generally ignored. Intrinsic losses are established for common bulk materials and can be easily measured in the case of pure liquids. The remaining three mechanisms, viscous, thermal and scattering, are considered important for acoustic spectroscopy in colloidal systems.

For high density-contrast systems ($\Delta\rho > 1$), like colloidal ceramic powders dispersed in an aqueous medium, the visco-inertial term dominates in the 1-100 MHz range, and thermal and scattering losses can be ignored. Scattering becomes increasingly important for particles larger than about 1 μm and at high frequencies, and is the dominant loss mechanism for particles larger than about 10 μm. For low $\Delta\rho$ materials, like emulsions and food colloids, thermal dissipation is typically the dominant loss mechanism (*20*), although in some cases (e.g., polymer latex) both visco-inertial and thermal losses may contribute significantly to attenuation. A large amount of research has been done on low $\Delta\rho$ systems, mainly because of their importance in the food industry, where acoustic techniques have been widely implemented (*21*). More recently, commercial devices have targeted high $\Delta\rho$ systems. An advantage of acoustic devices in general is their rather robust sensors and their capability to analyze flowing systems, making them potentially useful for on-line monitoring applications (*22*).

Attenuation and velocity can provide information about particle size and size distributions, dispersed phase concentrations, phase transitions, percolation thresholds, creaming and sedimentation, and other processes that modify the physical properties of the dispersed phase or lead to phase segregation. For particle sizing in high $\Delta\rho$ systems, the main input parameters are bulk density and volume fraction for each phase. For low $\Delta\rho$ systems, in addition to density and volume fraction, the thermophysical properties are required as inputs (i.e., thermal conductivity, specific heat capacity, thermal expansion coefficient) (*23*).

For particle sizing applications, attenuation is a more sensitive parameter than velocity. Currently used models for interpreting or predicting attenuation spectra in the MHz range are based on either "scattering" (*20, 21, 24*) or "coupled phase" approaches (*19, 23, 25*). In both treatments the medium is defined as an incompressible Newtonian fluid and the dispersed phase is regarded as an isolated spherical particle. The models differ primarily in the manner in which they account for higher order effects (e.g., particle-particle interactions). A comparison of models is complex and beyond the present scope.

It suffices to state that the efficacy of different models can depend on the specific nature of the test material, and that models are continually being improved.

The theory for dilute monodisperse systems, referred to as ECAH (after Epstein-Carhart-Allegra-Hawley) works fairly well at concentrations up to a few volume percent, and even higher in the case of thermal-loss dominated systems. More recent models permit measurements at higher concentrations and for polydisperse systems and multiple solid phases. In many cases, it is now possible to quantitatively determine particle size in suspensions with volume fractions of 30 % (*19, 23*). This is accomplished using either multiple scattering treatments or cell models that incorporate the effects of particle crowding on attenuation.

Acoustic Applications

The list of potential uses for ultrasonic spectroscopy in concentrated systems is long and varied. A number of applications of a practical nature are reviewed in several recently published works. McClements (*20*) and Povey (*21*) examine applications for emulsions and food colloids, including measurements of phase inversion, phase transition, coalescence, creaming, dispersed phase concentration and particle size. Dukhin and Goetz (*26*) discuss particle sizing in widely ranging material applications, from chemical mechanical polishing slurries to neoprene latex. In 1997 an international workshop on ultrasonic characterization of complex fluids was held at the National Institute of Standards and Technology. The proceedings (*27*) and conference report (*28*) for this workshop together provide a broad overview of the theory, technology and practice of radio-frequency acoustics in concentrated dispersed systems. To illustrate the advantages and limitations of ultrasonic spectroscopy, an example is given below for application to cement systems.

Acoustic Sensing of Cement Hydration

Cement is perhaps one of the more complex and difficult particle-fluid systems to characterize. Cement is concentrated, opaque, heterogeneous and chemically reactive. It has multiple solid phases, high ionic concentrations and is alkaline (> pH 12). A typical portland cement powder contains particle sizes spanning the range from colloidal dimensions to over 50 μm, with morphology that is often nonspherical. The setting of cement is the result of chemical and physical processes that take place between cement particles and water, and is largely driven by the hydration reactions of calcium and aluminum silicates. The principal hydration products are calcium silicate hydrates (CSH), calcium sulfoaluminate hydrates (ettringite) and $CaOH_2$. The exact composition of the hydration products is variable and depends on the starting composition of the cement and any chemical or mineral additives used.

As part of on-going studies at NIST, the acoustic attenuation spectra was measured in portland cement suspensions at volume fractions from 2 % to 20 % for periods of up to 7 hours after addition of water. This time period envelops the initial dissolution, subsequent dormant stage and acceleration period during which CSH and $CaOH_2$ are first precipitated. Initial studies indicate the high frequency portion of the attenuation spectra closely follows the progress of hydration reactions. Figure 3 shows the time-series evolution of attenuation spectra for a typical portland cement suspension in deionized water at a volume fraction of 5 %.

It is evident from this data that the low frequency attenuation is relatively insensitive to changes occurring during hydration. It has been found that the high frequency plateau above 65 MHz, at a fixed hydration time, is a linear function of the solids concentration (up to 20 %), indicating that the hydration

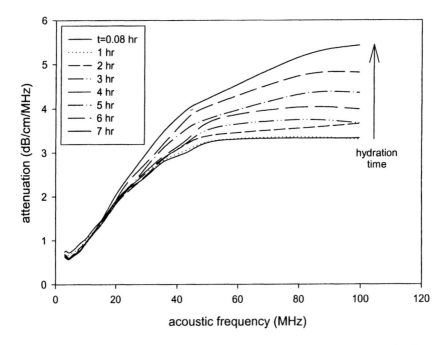

Figure 3. Acoustic spectra as a function of hydration time for a portland cement suspension at a volume fraction of 5 %.

process is largely independent of the solids concentration (at least during the early stages of hydration). Preliminary investigations demonstrate the efficacy of radio-frequency acoustics for in situ non-destructive monitoring of early stage hydration. The primary limitations of this method for cement applications are the lack of a clear understanding of the origins of the excess attenuation appearing at high frequencies, and the need for an appropriate theoretical model to predict it. This is the subject of on-going work in our laboratory, along with efforts to correlate acoustic data with other physical and chemical property measurements.

Electroacoustics

In electroacoustics one measures the response of charged particles to an applied acoustic or electric field. This response generates an oscillating double-layer polarization at the applied frequency, resulting from the relative motion between the particle and surrounding liquid (Figure 4). The corresponding signal gives the frequency-dependent dynamic mobility, which is the high-frequency counterpart to the d.c. mobility measured in a conventional microelectrophoresis experiment under dilute conditions. Three types of electroacoustic measurements are commercially available: electrokinetic sonic amplitude (ESA) where an a.c. electric field is applied and a sound wave is measured, colloid vibration potential (CVP) and colloid vibration current (CVI) where a sound

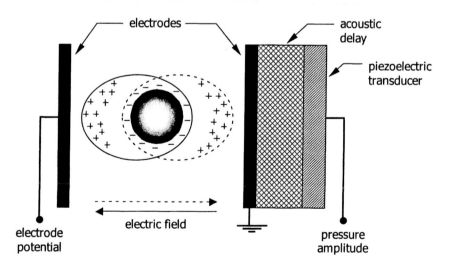

Figure 4. Schematic illustration of basic electroacoustic measurement design.

wave is applied and an a.c. potential or current, respectively, is measured. Instruments operate at either a fixed frequency or over a range of frequencies (electroacoustic spectroscopy). In the latter case, one can obtain particle size information by combining the magnitude and phase spectra using the theory of O'Brien (29). As particle size increases, particle motion lags behind the applied field, creating a phase shift and decreasing the magnitude. At very small sizes (sub-100 nm), the inertial forces become insignificant and the measured dynamic mobility is equivalent to the d.c. value, whereas at relatively large sizes (above 10 μm or so) particle motion is almost completely arrested due to inertia and the electroacoustic response is nearly zero. In between these limits, it is possible to determine both particle size and zeta potential simultaneously from the electroacoustic spectra, within the limits of the assumptions imposed by O'Brien's theory (e.g., thin double layer).

The electroacoustic effects, ESA and CVP are reciprocal in nature; that is, they are reciprocally related through the complex conductivity K^*. CVI, although it arises from the same response that produces CVP, is essentially equivalent to ESA, since it does not require determination of K^* (which must be known in order to calculate mobility from CVP measurements) .

At moderate concentrations (i.e., volume fractions between about 1 % and 10 %), the response is dependent on volume fraction, density contrast and dynamic mobility; other factors can be folded into an instrument constant that is determined using a known calibration "standard", such as Ludox silica or an ionic solution with a predictable response. There are certain drawbacks to using Ludox, but it at least allows for a common point of reference since it is widely available and commonly employed in commercial single frequency devices. There are several publications that address the issue of calibration, most notably by researchers at Kodak (30) and the University of Sydney (31). At higher particle concentrations, either semi-empirical (32) or theoretical extensions (e.g., cell models (33)) must be used to account for the nonlinear electroacoustic response due to particle interactions. Additionally, the acoustic impedance can no longer be assumed to be equivalent to the suspension medium (included in the instrument constant), and must therefore be measured separately (34).

In order to obtain an absolute value for zeta from an electroacoustic measurement, one must first correct for the inertial effect and this requires knowledge of the particle size in the test suspension (35), a parameter that will change during measurements if the suspension is unstable. In many cases, the particle size is not known and the resulting ESA or CVP signal can then, at best, provide an approximation of zeta. In some commercial devices, size (and therefore the magnitude of the inertial response) can be determined simultaneously (as in ESA spectroscopy) or in parallel (using for example acoustic attenuation spectroscopy), thereby providing a more accurate determination of zeta potential obtained under identical experimental conditions.

In many practical applications, knowledge of the absolute zeta potential is less important than tracking relative changes in zeta during a process. As such, electroacoustic titrations are commonly used to establish either the isoelectric point (IEP) or the effect of additives on the magnitude and polarity of the particle charge, without reference to an absolute value for zeta potential.

Other unique properties associated with electroacoustics include rapid measurements, robust sensors requiring no alignment and little maintenance, and less sensitivity to contamination relative to dilute measurements. Electroacoustics, like acoustics, are tolerant of flow conditions and are especially well suited for high-$\Delta\rho$ media. Although less established from a theoretical perspective, measurements can be performed in organic media; however, these applications have been reported in the literature to a lesser extent (36). Finally, the capability of measuring both size and zeta potential simultaneously in concentrated suspensions using multi-frequency methods is a valuable asset for researchers and industry alike.

In terms of limitations, size measurements are restricted for the most part to insulating particles. Semiconducting and conducting materials present problems, both theoretically and empirically. Non-insulating materials can also cause problems with measurement of mobility. For accurate inertial corrections, extremely broad or multimodal size distributions may prove difficult. Finally, one of the most persistent issues is due to the presence of electrolyte. High ionic concentrations can contribute significantly to the electroacoustic signal. This background signal must be subtracted to obtain the true colloid response (37). Certain ion pairs produce high signals and should be avoided; others, such as $NaNO_3$, produce a low response, and serve as good supporting electrolytes (38).

The main inputs needed for electroacoustic measurements are volume fraction and bulk density for the solid phase. In addition, the dielectric constant of the particulate phase is required for sizing using spectroscopic methods. As previously discussed, particle size must be known or determined along with the electroacoustic measurements in order to correct for inertial effects.

Although commercial devices have been available since the mid-1980s, the underlying technology and theory continue to evolve. For insight on the historical development of electroacoustics as a measurement technique for concentrated suspensions, see Zana and Yeager (39) and Babchin, Chow and Sawatsky (40). Publications by Hunter (34), Dukhin et al. (33) and Hackley and Texter (27) review more recent developments in this arena. A wide variety of applications are discussed, including ceramics, emulsions, pigments, cement, clays, fluorescent coatings, inks and toners.

For inorganic oxide-based suspensions, determination of the IEP is of both practical and fundamental importance, and electroacoustics are now utilized extensively for this purpose. A recent international joint effort involving the Japan Fine Ceramics Center, the Federal Institute for Materials Research and Testing in Berlin, and the National Institute of Standards and Technology, addressed technical and scientific issues relating to the application and standardization of methods for measuring the electrokinetic properties of moderately concentrated ceramic suspensions, including electroacoustics (*41*).

Summary

Recent developments in theory and instrumentation have led to improvements in the measurement of concentrated suspensions. A variety of methods are now available for characterizing these systems. Four key methods are presently reviewed: diffusing wave spectroscopy, ultra-small-angle x-ray scattering, acoustic spectroscopy and electroacoustics. Each method offers access to unique information on different properties of concentrated suspensions.

Acknowledgments

Acoustic attenuation spectra for portland cement measured by Lin-Sien Lum of NIST. Thanks to Andrew Allen of NIST for input on USAXS technique.

References

1. Berne, B. J.; Pecora, R. *Dynamic Light Scattering: With Applications to Chemistry, Biology, and Physics*; Dover Publications: Mineola, NY, 2000.
2. Pine, D. J.; Weitz, D. A.; Chaikin, P. M.; Herbolzheimer, E. *Phys. Rev. Lett.* **1988**, *60*, 1134-1137.
3. Scheffold, F. *J. Disp. Sci. Tech.* **2002** (in press).
4. Rojas-Ochoa, L. F.; Romer, S.; Scheffold, F.; Schurtenberger, P. **2002**, *Phys. Rev. E*, *65* (in press).
5. Zhu, J. X.; Durian, D. J.; Müller, J.; Weitz, D. A.; Pine, D. J. *Phys. Rev. Lett.* **1992**, *68*, 2559-2562.
6. Scheffold, F.; Skipetrov, S. E.; Romer, S.; Schurtenberger, P. *Phys. Rev. E* **2001**, *63*, 061404.
7. Romer, S.; Scheffold, F.; Schurtenberger, P. *Phys. Rev. Lett.* **2000**, *85*, 4980-4983.
8. Wyss, H. M.; Romer, S.; Scheffold, F.; Schurtenberger, P.; Gauckler, L. J. *J. Colloid Interface Sci.* **2001**, *240*, 89-97.
9. Mason, T. G.; Gang, H.; Weitz, D. A. *J. Opt. Soc. Am. A* **1997**, *14*, 139-149.
10. Dasgupta, B. R.; Tee, S.; Crocker, J. C.; Frisken, B. J.; Weitz, D. A. *Phys. Rev. E* **2002**, *65*, 051505.

11. Scheffold, F.; Romer, S.; Cardinaux, F.; Bissig, H.; Stradner, A.; Rojas-Ochoa, L. F.; Trappe, V.; Urban, C.; Skipetrov, S. E.; Cipelletti, L.; Schurtenberger, P. *Progr. Colloid Polym. Sci.* (in press).

12. Guinier, A.; Fournet, G. *Small-Angle Scattering of X-Rays*; John Wiley: New York, NY, 1955.

13. Long, G. G.; Krueger, S.; Jemian, P. R.; Black, D. R.; Burdette, H. E.; Cline, J. P.; Gerhardt, R. A. *J. Appl. Cryst.* **1990**, *23*, 535-544.

14. North, A. N.; Rigden, J. S.; Mackie, A. R. *Rev. Sci. Instrum.* **1992**, *63*, 1741-45.

15. Kerch, H. M.; Long, G. G.; Krueger, S.; Allen, A. J.; Gerhardt, R.; Cosandey, F. *J. Mater. Res.* **1999**, *14*, 1444-1454.

16. Ilavsky, J.; Allen, A. J.; Long, G. G.; Jemian, P. R. *Rev. Sci. Instrum.* **2002**, *73*, 1660-1662.

17. Long, G. G.; Jemian, P. R.; Weertman, J. R.; Black, D. R.; Burdette, H. E.; Spal, R. *J. Appl. Cryst.* **1991**, *24*, 30-37.

18. Allen, A. J.; Vertanessian, A.; Mayo, M. J. Nanoparticle agglomeration during drying studied by USAXS and FESEM. Presented at the Materials Research Society Fall Meeting, Boston, MA, 2001.

19. Dukhin, A. S.; Goetz, P. J. *Langmuir* **1996**, *12*, 4987-4997.

20. McClements, D. J. *Adv. Colloid Interface Sci.* **1991**, *37*, 33-72.

21. Povey, M. J. W. *Ultrasonic Techniques for Fluids Characterization*; Academic Press: London, UK, 1997.

22. Scott, D. M. In *Ultrasonic and Dielectric Characterization Techniques for Suspended Particulates*; Hackley, V. A. and Texter, J., Eds.; American Ceramic Society: Westerville, OH, 1998; pp 155-164.

23. Dukhin, A. S.; Goetz, P. J.; Hamlet, C. W. *Langmuir* **1996**, *12*, 4998-5004.

24. Riebel, U.; Löffler, F. *Part. Part. Syst. Charact.* **1989**, *6*, 135-143.

25. Harker, A. H.; Temple, J. A. G. *J. Phys. D: Appl. Phys.* **1988**, *21*, 1576-1588.

26. Dukhin, A. S.; Goetz, P. J. *Adv. Colloid Interface Sci.* **2001**, *92*, 73-102.

27. *Ultrasonic and Dielectric Characterization Techniques for Suspended Particulates*; Hackley, V. A.; Texter, J., Eds.; American Ceramic Society: Westerville, OH, 1998.

28. Hackley, V.A.; Texter, J. *J. Res. NIST* **1998**, *103*, 217-223.

29. O'Brien, R. W. *J. Fluid Mech.* **1990**, *212*, 81-93.

30. Klingbiel, R. T.; Coll, H.; James, R. O.; Texter, J. *Colloids Surf.* **1992**, *68*, 103-109.

31. O'Brien, R. W.; Cannon, D. W.; Rowlands, W. N. *J. Colloid Interface Sci.* **1995**, *173*, 406-418.

32. O'Brien, R. W.; Rowlands, W. N.; Hunter, R. J. In *Electroacoustics for Characterization of Particulates and Suspensions*; Malghan, S. B., Ed.;

Special Publication 856; National Institute of Standards and Technology: Gaithersburg, MD, 1993; pp 1-22.

33. Dukhin, A. S.; Shilov, V. N.; Ohshima, H.; Goetz, P. J. *Langmuir* **1999**, *15*, 6692-6706.

34. Hunter, R. J. *Colloids Surf. A: Physicochem. Eng. Aspects* **1998**, *141*, 37-65.

35. James, M.; Hunter, R. J.; O'Brien, R. W. *Langmuir* **1992**, *8*, 420-423.

36. Kornbrekke, R. E.; Morrison, I. D.; Oja, T. In *Electroacoustics for Characterization of Particulates and Suspensions*; Malghan, S. B., Ed.; Special Publication 856; National Institute of Standards and Technology: Gaithersburg, MD, 1993; pp 92-110.

37. Desai, F. N.; Hammad, H. R.; Hayes, K. F. *Langmuir* **1993**, *9*, 2888-2894.

38. Hackley, V. A.; Malghan, S. G. In *Electroacoustics for Characterization of Particulates and Suspensions*; Malghan, S. B., Ed.; Special Publication 856; National Institute of Standards and Technology: Gaithersburg, MD, 1993; pp 161-179.

39. Zana, R.; Yeager, E. In Modern Aspects of Electrochemistry; Bockris, J. O.; Conway, B. E.; White, R. E., Eds.; No 14; Plenum Press: New York, NY, 1982; pp 1-61.

40. Babchin, A. J.; Chow, R. S.; Sawatzky, R. P. *Adv. Colloid Interface Sci.* **1989**, *30*, 111-151.

41. Hackley, V.A.; Patton, J.; Lum, L.H.; Wäsche, R.; Abe, H.; Naito, M.; Hotta, Y.; Pendse, H. *J. Dispersion Sci. Tech.* **2002**, *23* (in press).

Chapter 6

Direct Measurement of Surface Forces for Nanoparticles

Jeong-Min Cho and Wolfgang M. Sigmund

Department of Materials Science and Engineering, University of Florida,
Gainesville, FL 32611–6400

The direct measurement of surface forces for nanoparticles is based on a modification of the colloid probe technique in an atomic force microscope. This technique has recently been developed by the authors. Here we present the impact of an adsorbed polymer layer on interparticle forces. Variation of the size of the colloid probe from micron to nano-size shows the impact of particle size on steric forces. Whereas interaction force for large particle decreased with a large amount of free polymer due to depletion effect, nanosize particles were less affected by depletion forces.

Introduction

The colloidal phenomena such as dispersion, agglomeration, consolidation and adhesion are controlled by the balance of forces acting on the particles [1]. Various forces participate in the net surface force of the system in the liquid medium such as van der Waals, electric double layer, solvation, hydrophobic, and polymeric forces. The dominance of a particular force decides the feature of the net surface force, either repulsive or attractive, and consecutively the stability and rheology of the given system. There have been several theories to explain the interaction between particle surfaces, one of which is DLVO (Derjaguin-Landau-Verwey-Overbeek) theory. DLVO is universal and mostly acceptable theory to understand the macroscopic colloidal behavior by interpreting the interparticular forces with van der Waals and electric double layer interactions [2]. However, there are a number of examples where DLVO theory does not adequately describe the measured interaction profiles. Because DLVO is based on a continuum approach, neglecting solvent effects such as ion-solvent interactions, surface-solvent interactions, and ion-ion correlation, non-DLVO forces occur when the property (e.g. density and mobility) and molecular structure of solvent near surfaces of particles are different from the bulk property [3]. For example, solvation or hydration force normally occurs between smooth surfaces such as mica at very short separation distances due to the overlap of two solvated surfaces, which arise from the specific solute-solvent and modified solvent-solvent interactions. Alternatively, attraction may occur due to the unfavorable structure of water (i.e. rearrangement of H-bond) near a hydrophobic interface [4]. Certainly neither all of these forces can occur simultaneously nor are independent of each other. They have different magnitudes and signs and one of their forces may become dominant as a function of separation distance and condition of the system.

Particle size is one of the critical factors in determining these interactions of particles. As particle size decreases, the interaction between the particles is more dependent on the surface properties because the ratio of surface area to bulk volume becomes large. When the particle size becomes close to the molecular size of the medium, the continuum theory of van der Waals forces or the mean field theory of double-layer forces should no longer be applied to describe the interaction between two surfaces [5].

van der Waals force is decided by Hamaker constant of the given material as follows:

$$F(H) = -\frac{AR}{12H^2} \quad (1)$$

, where A is Hamaker constant, R is particle radius, and H is separation distance. Hamaker constants can be calculated from Lifshitz theory, where the interacting bodies are treated as continua with certain dielectric properties. When it comes to nano-size particles, obtaining accurate dielectric spectral parameters over the entire frequency range is very difficult because quantum effects would influence the dielectric data of materials. In addition, local dielectric constant of the solvent may be changed because the mobility of the solvent molecules around small ions or particles is restricted in the solvation zone [4].

Polymers have been powerful additives as dispersants in colloidal industry to stabilize the colloidal particles and to achieve the maximum solids loading with steric repulsion between the molecules. They are adsorbed on the surfaces and form incompressible repulsive barriers. The properties such as solubility, molecular weight, and ionic charge developed in the liquid medium are critical factors to decide the proper dispersant in the given system. With micronsize particles, polymers are of size much smaller than colloidal particle and form certain conformations depending on the system condition. However, the interaction between a particle and a polymer of similar nanosize must be considered because they may form a different conformation on the surface from that of the large particles.

The atomic force microscopy (AFM) is the mostly used experimental tool to measure the surface forces. Many force measurements in AFM have been carried out using the colloidal probe technique developed by W. Decker et al [6]. However, utilization of AFM has several limitations for studying forces of nanosize particles. For example, a direct force measurement using a conventional probe technique in an AFM is limited to micron size colloidal particles due to the geometry and size of the tip. In order to overcome this size limit for the direct force measurement, a carbon nanotube (CNT) as micron-length spacer and as nanosized probe has been suggested [7, 8]. CNT can be a good alternative to nanoparticles for investigating the interaction force in AFM due to several advantages such as small diameter (2-20 nm), hemispherical end structure, low surface roughness, and high flexibility. CNTs as a STM or AFM probe generally has been developed for high-resolution imaging topography of a trenched sample [9]. Recently, several researches have imaged the surfaces in a liquid system [9, 10], but direct force measurement and theories of interaction between nanosize particles are not well developed and only a few attempts have been made using this approach.

In this paper the effect of the geometry of the AFM tip on the force measurement and interactions between the polymer-adsorbed particles depending on the size will be discussed.

Experimental Procedures

Force measurement was done with the multimode AFM (Nanoscope III, Digital Instrument, CA) using the contact mode Si tip (ESP, Digital Instrument, CA), whose spring constant and length of cantilever is 0.05 N/m and a 450 μm, respectively. All measurements were carried out in the liquid cell where the AFM tip was designed to submerge completely, by which capillary adhesion induced by a liquid meniscus between tip and sample can be reduced. 2 μm thick SiO_2 coated substrate by plasma enhanced chemical vapor deposited (PE-CVD) was supplied by Motorola and RMS roughness was less than 0.3 nm. For micronsized colloidal probe, a glassy carbon (Alfa Aeser, MA), which has a diameter of about 50 μm was attached on the cantilever using epoxy glue (UHU, Germany). For nano probe preparation, carbon soot (Alfa Aeser, MA) containing the carbonaceous material and multiwalled nanotubes (MWNT) was purified and many bundles of MWNT were manipulated to align on the temporary carbon tape. One MWNT bundle was then glued on the AFM using a small amount of epoxy glue in the inverted optical microscope (Carl Zeiss, Germany) with a method similar to that developed by H. Dai et al (9). After a long nanotube bundle was attached to the end, the AFM cantilever was heated to harden the epoxy glue at 200°C for 1 minute and air-cooled to room temperature. Then, the AFM tip was weakly flushed with DI water, leaving a strongly attached MWNT bundle with a single protruding nanotube. A branched polyethyleneimine (PEI, MW=70,000, 30 w/v%, Alfa Aeser, MA) was used as received for the steric force measurement. PEI has a high affinity to the silica and nanotube at high pH. The silica substrate was cleaned by ultrasonicating with the order of detergent, acetone, and ethanol for 30 minutes respectively and rinsed with DI water. The probe was cleaned with ethanol and massive water right before the measurement. 1 N HCl and 0.1 N NaOH were used for pH adjustment in 0.01 M NaCl ionic strength. After the colloidal probe and substrate were loaded in the AFM, the wanted solution was injected and allowed to equilibrate for 10 minutes. The force-separation profile was converted from the deflection signal and piezo scanner movement data, which was obtained by a Z-scan rate of 0.5 Hz.

Results and Discussion

Effect of the Geometry of AFM Tip

In theoretical calculations the standard AFM tip is very often considered as a sphere with a radius typically in the range of 5-100 nm [11, 12]. However, the standard AFM tip is of pyramidal shape and has irregular apex, which may contribute to the direct force measurement between the tip and substrate. Thus, tip can be treated as a cone with an angle of 2θ attached with a half sphere with radius R. The van der Waals force then can be described as [13]:

$$F(D)_{\text{Cone-plate}} = -\frac{A}{6}\frac{D^2(R+D)\sin^2\theta + D^2R\sin\theta + R^2(R+D)}{D^2(R+D)^2} \qquad (2)$$

Figure 1 shows how the side of the AFM tip will affect the total interaction force measurement depending on the angle between the vertical axis and side of the tip and the radius of curvature of the tip apex. As the angle increases, the contribution from the side of the tip increases. The difference between the interaction forces depending on the angle is not trivial as the radius of the tip apex decreases. However, when it comes to the nanotube probe, the angle between the wall and the tip apex can be considered as zero, minimizing the contribution from the side of the wall. The consequence is that the resultant force is very similar to the force calculated with the nanospheres as shown in figure 2.

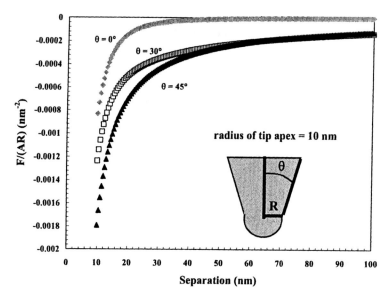

Figure 1. Calculated van der Waals interaction force between a cone shaped AFM tip and a flat plate depending on the angle θ.

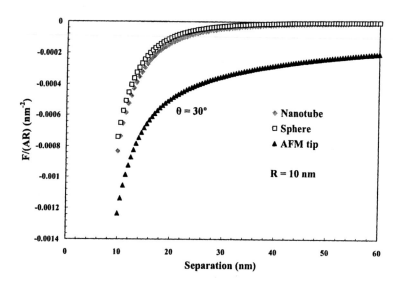

Figure 2. The calculated van der Waals interaction force between cone shaped AFM tip and flat plate comparing with the force between the sphere and flat plate.

Effect of the Colloidal Probe Size

Figure 3 shows the normalized force-separation distance plot measured with a glassy carbon sphere with the different amount of PEI. It was observed that the measured surface forces of micronsize colloidal probe were dependent on the polymer concentration. For the 0.06 wt% of PEI, incomplete coverage of the polymer layer seemed to take place considering the highest repulsive force occurred at a separation of 50 nm with 0.2 wt% of PEI. This repulsive barrier thickness is reasonable considering that the Flory radius (R_F) for a linear PEI is 31.2 nm. This fits well to the 50 nm size of the repulsive barrier thickness for the branched PEI. As the concentration of polymer was increased over 0.2 wt%, the interaction distance and repulsive forces were decreased. One feasible interpretation of this behavior may be the presence of free, non-adsorbed polymer. When a large concentration of polymer is not adsorbed on the surface and the separation between surfaces is reduced to a critical distance, attractive forces may occur by the depletion due to the induced osmotic pressure. Additionally, it cannot be excluded at this point that entanglement of polymers may occur and thus a reduction of the repulsive forces with an increased polymer concentration. A large adhesion force up to separation of 100 nm (not shown here) was observed when the tip was retracted from the substrate at pH 10.

Figure 4 shows the different force vs. separation profile depending on the existence of the free polymers in the solution. A non-depleted solution in the liquid cell was made by the following procedure. First, 0.2 wt% or 0.6 wt% PEI solutions were injected into the liquid cell, which were left for 10 minutes for polymers to adsorb on the surface of the sample, followed by replacing with the bulk solutions. The bulk solutions were polymer-free and had same pH and ionic background as the previously injected solution. For the 0.2 wt% solutions, both depleted and non-depleted solutions showed the same force-separation plot, which means all the polymers were adsorbed on the surface. On the contrary, much smaller repulsive forces were observed for the 0.6 wt% depleted solution than the non-depleted solution. A large amount of PEI, which was not adsorbed, will induce the depleted attractive force. The other possible reason of the decreased repulsive force is the changes in adsorbed conformation after free polymers were removed by washing procedure.

The force-separation distance profile with MWNT probe is shown in Figure 5. On the contrary to the micronsize particle, the same magnitudes of repulsive forces were observed even though the polymer concentration was varied. The interaction distance of a MWNT tip is of significantly shorter range than that of a micronsize glassy carbon tip. This behavior can be explained by the difference of probe size. The glassy carbon probe is much larger than the polymer allowing to interact with both tail and loop conformations of the adsorbed PEI and consequently measuring a far protruding layer. However, the nanosize MWNT is much smaller than the dimension of the polymer chain and the

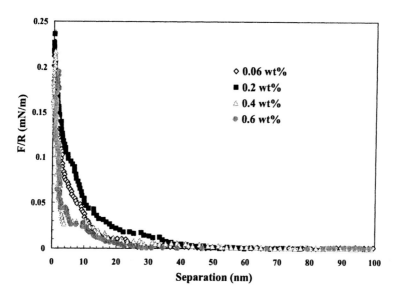

Figure 3. Force vs. separation plot between a glassy carbon and SiO₂ substrate.
Force dependence on the concentration of PEI was observed with the
microsized particle (pH 10, 0.01 M NaCl).

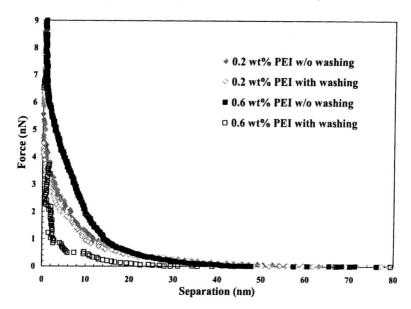

Figure 4. Force vs. separation plot between a glassy carbon probe and SiO₂
substrate in depleted and non-depleted solutions.

polymer tails easily diffuse out of the tips range when the nanosize tip approaches. Additionally, depletion effects cannot play a significant role since the size of the nano tip is too small compared to the polymer molecules, too. Thus, the measured forces are independent of the concentration of PEI. Even though different force behavior was observed for a large colloidal particle and nano particle, it is necessary to study further with molecules nonadsorbed to nanotube to distinguish the depletion force from the other forces. Depletion effect on the relative size of nonadsorbing particles for micronsize colloidal probe was recently studied with colloidal silica by Piech et al. They showed the decrease of depletion force with a large nonadsorbing particle [14].

Figure 6 shows the force-separation plot measured with a bare Si tip and silica substrate. The interaction distance is smaller than the case of micronsize colloidal probe but similar to that measured with the MWNT tip because the radius of curvature of tip ending is about 10 nm. However, the forces which result from the interaction between the side of the tip and substrate must be considered because the normal AFM tip is of a cone shape as explained in the above. Depletion effect was not observed with the bare silicon tip in the polymer concentration as high as 0.6 wt% to the contrary of the force measurement of glassy carbon sphere or MWNT probe. This may be from the different adsorption behavior of PEI on glassy carbon and Si i.e. glassy carbon has the less affinity of PEI on the surface.

Summary

It is expected that the use of the standard pyramidal AFM tip represents the forces occurring not only from the tip apex but also the side of the tip. This may be the source of the incorrect force measurement when simply assuming the radius of the tip apex to be that of counter nanoparticle. CNT in the conventional liquid mode of AFM opens the possibility to directly measure the interaction forces of true nanosize particles in various colloidal conditions by minimizing the contribution from the side of the tip. A nanosize MWNT probe clearly shows the independence of the concentration of polymer; whereas a micronsize glassy carbon sphere shows the larger interaction distance and the measured forces are functions of the concentration of polymer. Depletion effect was not observed in the MWNT probe and bare AFM tip because of the smaller size.

References

1. Horn, R. G., *J. Am. Ceram. Soc.* **1990**, *73*, 1117-1135.
2. Hunter, R. J. *Foundations of Colloidal Science;* Clarendon Press: Oxford, UK, 1989; Vol. 1.

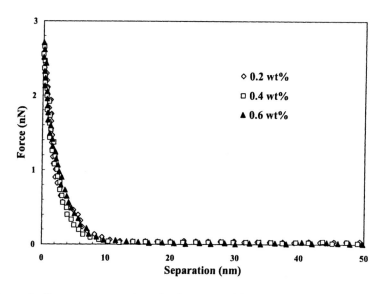

Figure 5. Force vs. separation plot between a MWNT probe and SiO₂ substrate. Measured forces were independent of the concentration of PEI (pH 10, 0.01 M NaCl).

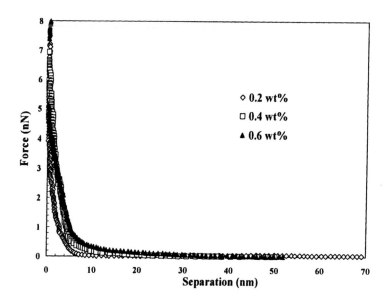

Figure 6. Force vs. separation plot between a bare Si tip and SiO₂ substrate (pH 10. 0.01 M NaCl)

3. Greathouse, J. A. and McQuarrie, D. A., *J. Colloid Interf. Sci.* **1996**, *181*, 319-325.
4. Israelachvili, J. N. *Intermolecular and Surfacd Forces;* Academic Press: San Diego, CA, 1991.
5. Mulvaney, P. In *Nanoparticles and Nanostructured Films*, Fendler, J.H. Ed.; Wiley-VCH, New York, 1998.
6. Ducker, W. A., Senden, T. J., and Pashley, R. M., *Nature* **1991**, *353,* 239-241.
7. Cho, J. M. and Sigmund, W. M., *J. Colloid Interf. Sci.* **2002,** *245,* 405-407.
8. Jarvis, S. P., Uchihashi, T., Ishida, T., Tokumoto, H., and Nakayama, Y., *J. Phys. Chem. B* **2000,** *104,* 6091-6094.
9. Dai, H. J., Hafner, J. H., Rinzler, A. G., Colbert, D. T., and Smalley, R. E., *Nature* **1996,** *384,* 147-150.
10. Moloni, K., Buss, M. R., and Andres, R. P., *Ultramicroscopy* **1999**, *80,* 237-246.
11. Argento, C. and French, R. H., *J. Appl. Phys.* **1996,** *80* 6081-6090.
12. Drummond, C. J. and Senden, T. J., *Colloid Surf. A-Physicochem. Eng. Asp.* **1994,** *87,* 217-234.
13. Pedersen, H. G., Ph.D. thesis, Technical University of Denmark, Copenhagen, Denmark, 1998.
14. Piech, M. and Walz, J. Y., *J. Colloid Interf. Sci.* 2002, 247, 327-341

Chapter 7

Scattering and Optical Microscopy Experiments of Polyelectrolyte Rodlike Cellulose Whiskers

M. Miriam de Souza Lima[1,2] and R. Borsali[1,*]

[1]LCPO-CNRS-ENSCPB, Bordeaux University, UMR 5629, 16, Avenue Pey
Berland, 33607 Pessac Cedex, France
[2]Department of Pharmacy and Pharmacology, Maringa State University,
CEP 87020–900, Maringa, Parana, Brazil
*Corresponding author: Borsali@enscpb.fr

Cellulose microcrystals can be produced by sulphuric acid hydrolysis
method from natural cellulose microfibrills obtained from different
sources. The dimensions for these rodlike polyelectrolytes depend on
their origin. These colloidal rod-like particles are electrostatically
stabilized in aqueous suspensions by the negative charges on their
surface. These charges result from the grafting of sulfate groups during
the acid hydrolysis process. In this work we have used ligth scattering
technique to investigate the structural ordering in these systems. This
long-range order was highlighted by the presence of scattering peaks as
function of the wavevector q at different whisker concentrations. The
electrostatic scattering maxima located at q_{max} observed in "salt-free"
whiskers suspensions scale roughly as $C^{1/2}$. As expected, at higher ionic
strength (added NaCl) the maxima disappear and these colloidal rods
behave as a neutral rod-like system. The measured effective diffusion
coefficient (dynamic behavior) $D(q)$ using dynamic light scattering is
found in excellent agreement with the static properties.

Cellulose is the most plentiful renewable biopolymer on earth. It is the main structural component in higher plants and some animals. Its basic chemical structure is formed by 1,4-joined β-D-glucopyranose residues. The large quantity of the hydroxyl groups present in the cellulose chain leads to a strong tendency to form intra and intermolecular hydrogen bonds. These interactions added to their β-1,4 configuration result in a rigid and linear polymer that contributes to the strong strength observed on the cellulose based organisms.[1,2]

Cellulose chains are biosynthesized as self-assembled microfibrils and exist in four different forms: cellulose I, II, III and IV polymorphous that can be interconverted by chemical and thermal process.[3,4,5] Several investigations have shown that the cellulose is constituted by two crystalline forms, cellulose I_α, present in algal and bacterial organisms and cellulose I_β present in higher plants. Due to its crystalline structure and interesting properties the cellulose is widely used in paper, food, additives, optical and pharmaceutical industries.

The presence of a crystalline and amorphous regions on the cellulose microfibrils result in an irregular surface that is susceptible to acid hydrolitic cleavage. The controled sulfuric acid hydrolysis on the cellulose microfibrils produces rod-like microcrystals (commonly called cellulose whiskers) that are negatively charged on their surface. After the hydrolysis process the colloidal whisker suspensions neither precipitate nor flocculate due the electrostatic repulsions between the particles.[6-11] The electrostatic interactions between these rod-like polyelectrolytes whiskers are at the origin of very interesting properties.[8]

The typical dimensions (length L and diameter d) of the cellulose rods depends on their cellulose origin. The cotton cellulose whiskers have $L = 200$-300 nm and $d = 15$ nm while the tunicate whiskers have $L = 2000$ and $d = 15$-20 nm.[7,8] The large axis ratio (L/d) in these rods is the more important parameter that determine their ordering behavior as well as their liquid crystal properties.[9-12]

In aqueous dilute suspensions the cellulose rods are randomly oriented (isotropic phase). As the concentration is increased the cellulose rods self-align along a vector director forming a nematic phase. Above a critical value of the concentration, these rods display a cholesteric phase where the ordering is defined by the helicoidal stacking of the nematic planes.[9-13] Macroscopically these properties can be evidenced using polarised microscopy as illustrated in figure 1. Figure 1a represents a dilute suspension of cellulose rods where the presence of some birefringent droplets indicate the initial organization stage of these rods.

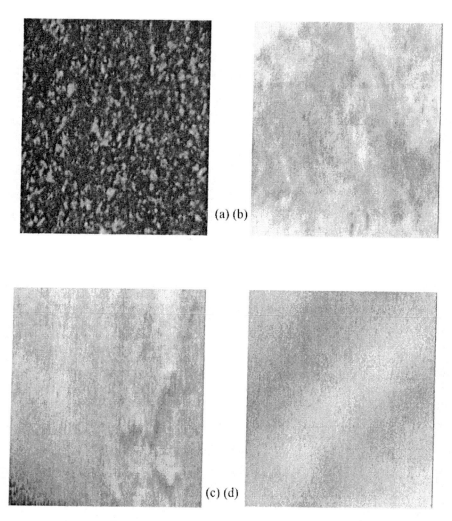

(a) (b)

(c) (d)

Figure 1 - Cellulose whiskers observed under crossed polarized microscopy in a dilute suspension (isotropic phase) (a), at concentrated regime (b) and at higher concentrations (cholesteric phase, c and d).

Increasing the concentration, by water evaporation for instance, leads to the existence of different colored domains reflecting the self-alignment of the whisker rods (figure 1b).

At higher concentration the presence of well defined color bands suggest the formation of cholesteric phases (figure 1c and 1d).

Due to their interesting optical properties these cellulose whiskers and their capability to form colored films with chiral nematic structure are used to make novel pigments for coating and inks.[14]

Experimental observations have shown that the alignment of these rods that can be done by shearing the systems and permited at the same time to better understand the ordering mechanism observed in these systems. Based on several experimental observations, it was proposed that the existence of a chiral twist agent with the ionic layer around each rod is at the origin of their helicoidal packed structure. This result was shown to be dependent on the axis ratio (L/d) of these whiskers and of the ionic strength.[12,15,16] As was also observed using small angle synchrotron radiation scattering under shear, the result suggests the existence of planar domains of randomly oriented cellulose whiskers which align at low shear rates and are broken up at higher shear displaying individual alignment of rod.[17]

The very easy alignment of these whiskers in the same direction (nematic ordering) make them a promising tensile material in the fiber industry. These properties are at to origine of their large use for instance as nanocomposites in polymeric matrix.[7,18-20]

To highlight the polyelectrolytic nature of the cellulose whiskers and therefore the ordered structure in these systems, we have performed light scattering experiments on aqueous whisker suspensions obtained from tunicate.[8] For this purpose we have fractionated these rods in order to decrease their polydispersity in length. The experiments were carried out at low concentration C *(isotropic phase)* in "salt-free" conditions and at several ionic strengths (by addition of NaCl) using static light scattering (*SLS*) and dynamic light scattering (*DLS*).

Experimental Section

1. Sample preparation – Stable aqueous suspensions of sulphuric acid-hydrolyzed cellulose whiskers were obtained from tunicate (*Microscosmus fulcatus*) mantles as previously described.[8] To reduce the polydispersity in length of these samples, we have used a fractionation method using ultracentrifugation process with a saccharose gradient as previuosly described.[8] The samples were concentrated, purified and checked by Transmission Electron Microscopy (TEM). The results showed that we have obtained different size distributions of cellulose whiskers as function of their length L. This fractionation method was effective and indeed narrowed the width of the whisker distribution length.

2. Equipment and data analysis – The ligth scattering experiments were carried out using the ALV (Langen-FRG) equipment with an automatic goniometer table, a digital rate meter and a temperature control to stabilize the temperature of the sample cell at $25\pm0.1°$C. The scattered light of a vertically polarized $\lambda_0=4880$ Å argon laser (Spectra-Physics 2020, 3W, operating around 0.3 W) was measured at different angles in the range of $20°$-$150°$ corresponding to 0.6 x $10^{-3}<q/Å<3.3 \times 10^{-3}$ where $q=(4 \pi n/\lambda_0) \sin (\theta /2)$, θ the scattering angle, n the refractive index of the medium ($n=1.33$). The reduced elastic scattering $I(q)/kC$, with $K=4\pi^2 n_0^2 (dn/dc)^2 (I_0^{90°}/R^{90°})/\lambda_0^4 N_A$, was measured in steps of $2°$ in the scattering angle, where n_0 is the refractive index of the standard (toluene), $I_0^{90°}$ and $R^{90°}$ are respectively the intensity and the Rayleigh ratio of the standard at $\theta=90°$, the increment of refractive index, dn/dc is 0.103, C the whisker concentration, expressed in g/cm^3 and $I(q)$ the intensity scattered by the sample. All elastic intensities were calculated according to standard procedures using toluene as standard with known of absolute scattering intensity. The static structure factor $S(q)$ was calculated from $I(q)=P(q) S(q)$ where $P(q)$ denotes the form factor of the rod-like particles in the isotropic phase.[21] For the dynamic properties, the experiments were carried out in steps of $2°$ in the scattering angle. The ALV5000 autocorrelator (ALV, FRG) was used to compute the autocorrelation functions $I(q,t)$ from the scattered intensity data. The autocorrelation functions of the scattered intensity, deduced from the Siegert relation, were analyzed by means of cumulant method to give the effective diffusion coefficient $D_{eff}(q)=\Gamma(q)/q^2$ as a function of the wavevector q.

The light scattering experiments were performed on the obtained fractionated whiskers particles whose average length L is about 9000 Å and a diameter d about 150 Å. All the samples were washed by centrifugation using milli-Q water (conductivity = 18.2 MΩ.cm). After the washing steps they were dialyzed against milli-Q water and put into an ion-exchange resin for few days prior to measurements. The resin was removed by centrifugation. For all studied concentrations in "salt-free" conditions, the stock solution was diluted with milli-Q water. The samples were filtered using an ion-exchange resin column and all the experiments were made in the presence of ion-exchange resin in the light scattering cuvette.

Results and Discussion
1. Static behavior
The scattering behavior of polyelectrolyte systems in "salt-free" conditions is characterized by the existence of a scattering peak at certain wavevector value (q_{max}) that depends on the concentration.[22-26] As observed for other systems q_{max} scales with the polyelectrolyte concentration as $C^{1/2}$ in semi-dilute solutions and $C^{1/3}$ in dilute regime. The existence of the scattering peak is due to the preferential distance

electrostatically imposed between the charged particles resulting in pseudo-organized domains displaying cubic arrangement in dilute regime ($q_{max} \sim C^{1/3}$) and cylindrical/hexagonal packing in semi-dilute regime ($q_{max} \sim C^{1/2}$).

For spherical particles that interact via electrostatic interactions the structure factor S(q) can be calculated from the scattered intensity I(q) through the relation I(q)=S(q)P(q) where P(q) is the factor form factor of spherical particles. In the case of rod-like particles, the relation I(q)=S(q)P(q) is in general not valid because of the correlation between the mutual relative orientations of the rod-like particles and their relative spatial positions. Indeed for isolated pairs of rigid charged rods, mutually orthogonal orientations become strongly favored upon close approach, so orientation is strongly correlated with their relative spatial position.

However the use of the relation I(q)=S(q)P(q) in the case of rod-like particles may still be valid in two cases: i) At high concentrations of charged rods, the contribution of higher order clusters (in the virial sense) dominates that of the pair clusters, and the neighboring rods within a cluster adopt a parallel orientations. ii) For long rigid rods most of the scattered light at high q-values, comes from rods with orientations nearly perpendicular to the scattering vector q. Hence, the scattering will be dominated by those clusters with rods oriented nearly perpendicular to q (the scattering coming from orthogonal neighbors is negligible). When q and L are sufficiently large, such as qL>>1.0, P(q) selects predominantly rods that are oriented perpendicular to q and as a consequence it varies rather slowly with q, as reported by Wilcoxon and Schurr.[27] Under this circumstance, when only rods nearly perpendicular to q scatter significant light and P(q) is roughly constant, and the neighboring rods have all nearly the same orientation in any case, the relation I(q)=S(q)P(q) should again be a good approximation.

In our system, although the whisker concentration is not sufficiently high, the relation I(q)=S(q)P(q) is still valid due the magnitude of the wavevector q and the whisker length, L. For this system the quantity qL, varies from qL=5.4 to qL=29.7. Based on these considerations, S(q) can be deduced from the ratio I(q)/P(q) where P(q) is the form factor of the rod of length L given by:

$$P_{rod}(q) = \frac{2}{qL} \int_0^{qL} \frac{\sin x}{x} dx - \left(\frac{\sin\left(\frac{qL}{2}\right)}{\frac{qL}{2}} \right)^2$$

(1)

At low ionic strength ("salt-free" suspensions) the cellulose whiskers show broad and well defined scattering peaks in a structure factor S(q) as a function of the wavevector q. Figure 2 presents S(q)=I(q)/P(q) as a function of q (C=1.04 x 10^{-2}

g/cm³) where a second and a third maximum are evidenced, suggesting the strong electrostatic interactions and the long-range-order in these systems.

At relatively high ionic strength obtained by the addition of NaCl(10^{-5} Mol/L), the electrostatic interactions are screened out and the rods behave as neutral systems where I(q) is a decreasing function of q. Indeed the scattering peak disappear (at 10^{-5} Mol/L) confirming the strong electrostatic interactions between the cellulose rods in salt-free suspensions.

The scattering wavevector of the first peak maxima scale as $C^{1/2}$. This exponent is close to the theoretical prediction of 0.5 for a geometrical model of cylindrically packed rods.[28-30] This behavior is illustrated in figure 3.

DLS experiments were also performed on "salt-free" whisker suspensions at the same concentration namely of C=1.04 x 10^{-2} g/cm. The full homodyne autocorrelation functions of the scattered intensity were obtained using the ALV-5000 autocorrelator. The measured intensity-intensity time correlation function is related to the electric field correlation function by Siegert relation:[31] $G^{(2)}(t)=B(1+\beta|G^{(1)}(t)|^2)$, where B is a baseline and β is the spatial coherence factor depending upon the geometry of the detection and the ratio of the intensity scattered by the particle to that scattered by the solvent. Generally, $G^{(1)}(t)$ may be expressed by a continuous distribution of decays: $G^{(1)}(t)=\int A(\Gamma)\exp - (\Gamma t) d \Gamma$ where $G^{(1)}(t)$ is the Laplace transform of the decay rate distribution function $A(\Gamma)$. This quantity, $A(\Gamma)$, gives the relative intensity of scattered light with decay constant Γ and is a function of the number and size of the scatteres.[21] For a dilute solution of monodisperse particles (spherical or coiled polymers) undergoing a Brownian motion, $G(\Gamma)$ may be represented by a single relaxation: $G^{(1)}(t) = \exp (-q^2Dt)$ where D is the translational diffusion coefficient.

Figure 4 shows the angular variation of the reciprocal effective diffusion coefficient $D(q)^{-1}=[\Gamma(q)/q^2]^{-1}$ deduced from the standard second order cumulant analysis of the autocorrelation functions:

$$\ln |S(q,t)| = - <\Gamma> \tau + \frac{\mu_2}{2!} \tau^2 - \frac{\mu_3}{3!} \tau^3 + ... \qquad (2)$$

Both I(q) and $[\Gamma(q)/q^2]^{-1}\equiv D^{-1}(q)$ have qualitatively the same shape and approximately the maxima at the same q-position. This result shows that the mobility M(q) is independent of q according to the general Ferrel-Kawasaki expression for the apparent diffusion coefficient $\Gamma(q)/q^2 \approx M(q)/S(q) \equiv D(q)$, as has often been observed.[25,27,32] This non-dependence of M(q) vs q can be explained as follows: The q-dependence of M(q) is due to the fact that at very small q orientational averaging yields the rotationally averaged translational diffusion coefficient $D_0=(1/3)(D_{||}$ $+2D_{\perp})$, where $D_{||}$ is the parallel diffusion coefficient and $D_{\perp}$ is the perpendicular

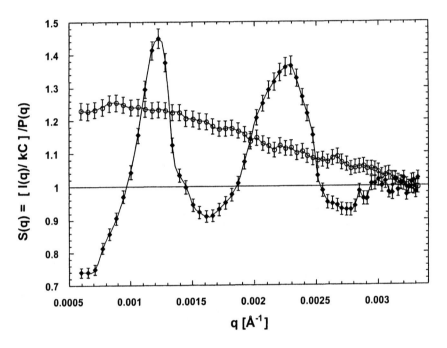

Figure 2 - Structure factor S(q) versus q from the tunicate whisker suspensions at concentrations of 1.04 x 10⁻² g/cm³ in in "salt-free" conditions (◆) and in presence of 10⁻⁵ Mol/L of NaCl (◇).

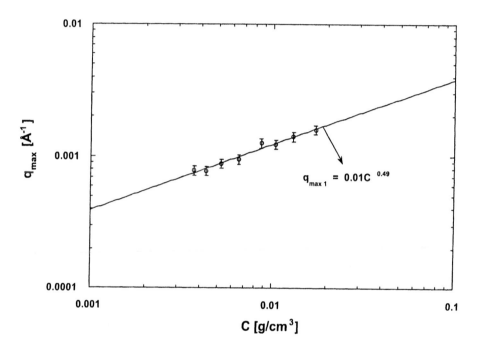

Figure 3 - Variation of Log q_{max} as a function of Log C. The q_{max} values corresponds to the first maxima (O). Although not plotted here, the second and the third scattering peaks scales as $C^{0.5}$.

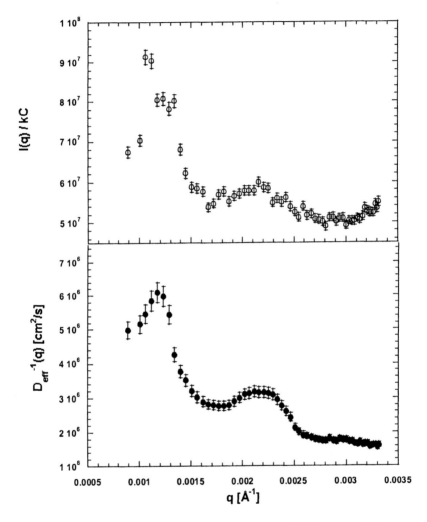

Figure 4 - Angular variation of the reciprocal diffusion coefficient $D(q)^{-1}=[\Gamma(q)/q^{2}]^{-1}$ (■) determined from the first order cumulant analysis of the autocorrelation functions DLS compared to $I(q)/kC$ (□) at $C=1.04 \times 10^{-2}$ g/cm^{3}, measured by static light scattering.

diffusion coefficient to the rod axis. At high q, the diffusion is dominated by $D_{\perp}$ and D_{rot} which are the only motions capable of modulating the intensity when the rods that dominate the scattering are oriented nearly perpendicular to q. In the present case, even for the smaller q-values, qL is still so large that $D_{\perp}$ and D_{rot} dominate the scattering, in which case, M(q) is relatively insensitive to changes in q.

Depending on the range of concentration (dilute or semi-dilute) the difference between the detailed shapes of the peaks in $[\Gamma(q)/q^2]^{-1}$ versus q is often attributed to hydrodynamic effects.[33]

Conclusion

The static and dynamic properties of rod-like polyelectrolyte cellulose whiskers have been investigated using elastic and quasi-elastic light scattering technique. Our light scattering results showed a clear evidence of several pronounced scattering peaks as a function of wavevector q. These peaks are due to the strong electrostatic interactions between the whiskes rods and disappear as the ionic strength is increased. The presence of a second and third peaks highligth the long range electrostatic order that exists in these whisker suspensions.

A scaling law $q_{max} \cong C^{1/2}$, found in our experiments suggests a cylindrical/hexagonal packing of these rods. As expected, the effective diffusion coefficient $D_{eff}(q)=\Gamma(q)/q^2$ is found inversely proportional to the structure factor S(q).

Acknowledgments
M. Miriam de Souza Lima thanks the financial support from CAPES – Brazil during her Phd thesis in France.

References

1. Coffey, D. G.; Bell, D. A. and Henderson, A. **1995**, Cellulose and Cellulose derivatives, In: *Food Polysaccharides and their Applications.* Alistar M. Stephen (Ed), New York, 124.
2. Whistler, R. L. and BeMiller, J. M. **1997**, Cellulosics In: *Carbohydrate Chemistry for Food Scientists,* American Association of Cereal Chemists, Inc, Minnesota, 7, 155.
3. Woodcock, C. and Sarko, A. *Macromolecules,* **1980**, 13, 1183.
4. Sugiyama, J., Persson, J. and Chanzy, H. *Macromolecules,* **1991**, 24, 2461.
5. Atalla, R. H. **1989**, In: *Biomedical and Biotechnological Advances in Industrial Polysaccharides,* Crescenzi, V., Dea, I. M. C., Paoletti, S., Stivala, S. S. and Sutherland, I. W. (Eds), Gordon & Breach, New York, 399.

6. Wise, L. E., Murphy, M. and d'Addiecco, A. A. *Pap. Trade J.*, **1946**, 122, 35.
7. Favier, V.; Chanzy, H.; Cavaillé, J. Y. *Macromolecules*, **1995**, 28, 6365.
8. De Souza Lima, M. M. and Borsali, R. *Langmuir*, **2002**, 18, 992.
9. Revol, J. F.and Marchessault, R. H. *Int. J. Biol. Macromol.*, **1993**, 15, 329.
10. Revol, J. F.; Godbout, L. ; Dong, G, X. M. ; Gray, D. G. ; Chanzy, H. and Maret, G. *Liquid crystals*, **1994**, 16, 127.
11. Revol, J. F., Orts, W. J., Godbout, L and Marchessault, R. H. *Polymeric Materials Science and Engineering*, **1994**, 71, 334.
12. Dong, X. M. and Gray, D. G. *Langmuir*, **1997**, 13, 11, 3029.
13. Araki, J and Kuga, S. *Langmuir*, **2001**,12, 449.
14. Revol, J.-F.; Godbout, L.; Gray, D. G. *J. of Pulp and Paper Sci.*, **1998**, 24, 5, 146.
15. Orts, W. J., Godbout, L. Marchessault, R. H. and Revol, J. F. *Macromolecules*, **1998**, 31, 5717.
16. Inatomi, S., Jinbo, Y., Sato, T. and Teramoto, A. *Macromolecules,* **1992**, 25, 5013.
17. Ebeling, T.; Borsali, R.; Paillet, M.; Diat, O.; Cavaille, J.Y.; Chanzy, H.; Dufresne, A. *Langmuir*, **1999**, 19, 6123.
18. Dubief, D., Samain, E. and Dufresne, A. *Macromolecules*, **1999**, 32, 5765.
19. Chazeau, L., Paillet, M. and Cavaillé, J. Y. *J. of Polymer Sci.: Part B: Polymer Physics*, **1999**, 37, 2151.
20. Chazeau, L., Cavaillé, J. Y. and Terech, P. *Polymer*, **1999**, 40, 5333.
21. Berne, B. J. and Pecora, R. **1976** *Dynamic Light Scattering*. New York : Wiley Interscience.
22. Maier, E. E.; Krause, R.; Deggelmann, M.; Hagenbüchle, M; Weber, R. *Macromolecules.*, **1992**, 21, 1125.
23. Graf. C.; Deggelmann, M.; Hagenbüchle, M; Kramer, H.; Krause, R.; Martin, C. and Weber, R. *J. Chem. Phys.*, **1991**, 95 (9), 12, 6284.
24. Schulz, S.; Maier, E. E.; Weber, J. *J. Chem. Phys.*, **1989**, 90, 1, 7.
25. Xiao, L. and Reed, W . F. *J. Chem. Phys.*, **1991**, 94, 4568.
26. Borsali, R. **2002**, *Polyelectrolytes, their characterization, Polyelectrolyte Solution, In: Handbook of Polylectrolytes and their Applications*, Tripathy, S. K., Kumar, J. and Nalwa, H. S. (Eds) 9, 1.
27. Wilcoxon, J. Schurr, J. M. *Biopolym.*, **1993**, 22, 849.
28. Benmouna, M.; Weill, G.; Benoit, H. and Akcasu, Z. *J. Phys. France*, **1982**, 43, 1679.
29. De Gennes, P. G. **1979** *Scaling Concepts in Polymer Physics*, Cornell University Press, Ithaca.
30.Hayter, J.; Jannink, G.; Brochard, F. and De Gennes, P. G. J. Phys. Lett. France, 1980, 41, 451.
31. Siegert, A. J. *MIT Rad. Lab. Rep.* **1943** (n. 465).
32. Driford, M. and Dalbiez, J. P. *J. Phys. Chem.*, **1984**, 88, 5368.
33. Van Vinckle, D. H., Murray, C. A. *Phys. Rev. A.*, **1986**, 36, 1, 562.

Chapter 8

Ultrasound for Characterizing Colloids

Andrei S. Dukhin and Philip J. Goetz

Dispersion Technology Inc, 364 Adams Street, Bedford Hills, NY 10507

Ultrasound provides unique means for characterizing concentrated colloids. For a long time this simple fact has been practically completely ignored in the Colloid Science. Recently, situation changed, more and more groups around the world apply ultrasound for characterizing various concentrated colloids. The current state of this field is described in details in the book with the title that is identical to the title of this short review [1]. The purpose of this review is to place ultrasound based techniques within a scope of traditional methods of colloids characterization.

There are two major scientific discipline , Acoustics on one hand and Colloid Science on the other, that are involved in the development of the ultrasound methods for concentrated colloids characterization. It is a rather curious situation that, historically, there has been little real communication between disciples of these two fields. Although there is a large body of literature devoted to ultrasound phenomena in colloids, mostly from the perspective of scientists from the field of Acoustics, there is little recognition that such phenomena may be of real importance for both the development, and application, of Colloid Science. From the other side, colloid scientists have not embraced acoustics as an important tool for characterizing colloids. The lack of any serious dialog between these scientific fields is perhaps best illustrated by the fact that there are no references to ultrasound or Acoustics in the major handbooks on Colloid and Interface Science [2,3], nor any reference to colloids in handbooks on acoustics [3,4].

One might ask "Perhaps this link does not exist because it is not important to either discipline?" In order to answer this question, let us consider the potential place of Acoustics within an overall framework of Colloid Science. For this purpose, it is helpful to classify non-equilibrium colloidal phenomena in two dimensions; the first determined by whether the relevant disturbances are electrical, mechanical, or electro- mechanical in nature, and the second based on whether the time domain of that disturbance can be described as stationary, low frequency, or high frequency. Table 1 illustrates this classification of major colloidal phenomena. The low and high frequency ranges are separated based on the relationship between either the electric or mechanical wavelength λ, and some system dimension L.

Clearly, light scattering represents electrical phenomena in colloids at high frequency (the wavelength of light is certainly smaller than the system dimension). There was, however, no mention in colloid textbooks, until very recently, of any mechanical or electro-mechanical phenomena in the region where the mechanical or electrical wavelength is shorter than the system dimension. This would appear to leave two empty spaces in Table 1. Such mechanical wavelengths are produced by "Sound" or,

Table 1. Non-equilibrium colloidal phenomena

	Electrical nature	Mechanical nature	Electro-mechanical	Mechano-electric
Stationary	Conductivity Surface conductivity.	Viscosity, Stationary colloidal hydrodynamics, Osmosis, Capillary flow.	Electrophoresis, Electroosmosis, Electro-viscosity	Sedimentation current/potential Streaming current/potential,
Low frequency (λ>L)	Dielectric spectroscopy	Oscillatory rheology.	Electro-rotation, Dielectrophoresis Electrophoretic light scattering	
High frequency (λ<L)	Optical scattering, X-ray spectroscopy	Empty? Acoustics!	Empty? Electroacoustics! ElectricSonic Amplitude	Empty? Electroacoustics! Colloid Vibration Current/Potential

when the frequency exceeds our hearing limit of 20 KHz, by "Ultrasound". For reference, ultrasound wavelengths lie in the range from 10 microns to 1 mm, whereas the system dimension is usually in the range of centimeters. For this reason, we consider ultrasound related effects to lie within the high frequency range. One of the empty spaces can be filled by acoustic measurements at ultrasound frequencies, which characterize colloidal phenomena of a mechanical nature at high frequency. The second empty space can be filled by electroacoustic measurements, which allow us to characterize electro-mechanical phenomena at high frequency. This book will help fill these gaps and demonstrate that acoustics (and electroacoustics) and can bring much useful knowledge to Colloid Science. As an aside, we do not consider here the use of high power ultrasound for modifying colloidal systems, just the use of low power sound as a non-invasive investigation tool that has very unique capabilities.

There are several questions that one might ask when starting dealing with ultrasound. We think it is important to deal with these questions right away, at least giving some preliminary answers, which are clarified and expanded in the book [1]. Here are these questions and the short answers.

Why should one care about Acoustics if generations of colloid scientists worked successfully without it?

While it may be true at present that the usefulness of Acoustics is not widely understood, it seems that earlier generations had a somewhat better appreciation. Many well-known scientists applied Acoustics to colloidal systems, as will be described in a detailed historical overview in the next section. Briefly, we can mention the names of Stokes, Rayleigh, Maxwell, Henry, Tyndall, Reynolds, and Debye, all of whom considered acoustic phenomena in colloids as deserving of their attention. The first colloid-related acoustic effect to be studied was the propagation of sound through fog; contributions by Henry, Tyndall and Reynolds made more than century ago between 1870-80. Another interesting, but not so well known fact, is that Lord Rayleigh, the first author of a scattering theory, entitled his major books "Theory of Sound". He developed the mathematics of scattering theory mostly for sound, not for light as is often assumed by those not so familiar with the history of Colloid Science. In fact, the main reference to light in his work was a paragraph or two on "why the sky is blue".

If Acoustics is so important, why has it remained almost unknown in Colloid Science for such a long time?

We think that the failure to exploit acoustic methods might be explained by a combination of factors: the advent of the laser as a convenient source of monochromatic light, technical problems with generating monochromatic sound beams within a wide frequency range, the mathematical complexity of the theory, and complex statistical analysis of the raw data. In addition, acoustics is more dependent on mathematical calculations than other traditional instrumental techniques. Many of these problems have now been solved mostly due to the advent of fast computers and the development of new theoretical approaches. As a result there are a number commercially available instruments utilizing ultrasound for characterizing colloids, produced by Matec, Malvern, Sympatec, Colloidal Dynamics, and Dispersion Technology.

What information does ultrasound based instruments yield?

For colloidal systems, ultrasound provides information on three important areas of particle characterization: Particle sizing, Rheology, and Electrokinetics.

In addition, ultrasound can be used as a tool for characterizing properties of pure liquids and dissolved species like ions or molecules, but we will cover this aspect only briefly.

An Acoustic spectrometer may measure the attenuation of ultrasound, the propagation velocity of this sound, and/or the acoustic impedance, in any combination depending on the instrument design. The measured acoustic properties contain information about the particle size distribution, and volume fraction, as well as structural and thermodynamic properties of the colloid. One can extract this information by applying the appropriate theory in combination with a certain set of a'priori known parameters. Hence, an Acoustic spectrometer is not simply a particle-sizing instrument. By applying sound we apply stress to the colloid and consequently the response can be interpreted in rheological terms.

In addition to acoustics there is one more ultrasound-based technique, which is called Electroacoustics. The Electroacoustic phenomenon, first predicted by Debye in 1933, results from coupling between acoustic and electric fields. There are two ways to produce such an Electroacoustic phenomenon depending on which field is the driving force. When the driving force is the electric field and we observe an acoustic response we speak of ElectroSonic Amplitude (ESA). Alternatively, when the driving force is the acoustic wave we speak, instead, of the Colloid Vibration Potential (CVP) if we observe an open circuit potential, or a Colloid Vibration Current (CVI) if we observe a short circuit current. Such electroacoustic techniques yield information about the electrical properties of colloids. In principle, it can also be used for particle sizing.

Where can one apply ultrasound?

The following list gives some idea of the existing applications for which the ultrasound based characterization technique is appropriate:

Aggregative stability, Cement slurries, Ceramics, Chemical-Mechanical Polishing, Coal slurries, Coatings, Cosmetic emulsions, Environmental protection, Flotation, Ore enrichment, Food products, Latex, Emulsions and micro emulsions, Mixed dispersions, Nanosized dispersions, Non-aqueous dispersions, Paints, Photo Materials.

This list is not complete. There is a table in the book [1] that summarizes all experimental works currently known to us.

What are the advantages of ultrasound over traditional characterization techniques?

There are so many advantages of ultrasound. The last section of this short review is devoted to describing the relationship between ultrasound based and traditional colloidal characterization techniques.

Historical overview

The roots of our current understanding of sound go back more than 300 years to the first theory for calculating sound speed suggested by Newton [6]. Newton's work is still interesting for us today because it illustrates the importance of thermodynamic considerations in trying to adequately describe ultrasound phenomena. Newton assumed that sound propagates while maintaining a constant temperature, i.e. an isothermal case. Laplace later corrected this misunderstanding by showing that it was actually adiabatic in nature [6].

This thermodynamic aspect of sound provides a good example of the importance of keeping a historical perspective. At least twice during the past 200 years the thermodynamic contribution to various sound-related phenomena was initially neglected, and only later found to be quite important. This thermodynamic neglect happened first in the 19th century, when

Stokes's purely hydrodynamic theory for sound attenuation [9, 10] was later corrected by Kirchhoff [7, 8]. Then again, in the 20th century, Sewell's hydrodynamic theory for sound absorption in heterogeneous media [12] was later extended by Isakovich [11] by the introduction of a mechanism for thermal losses.

We have now a very similar situation concerning electroacoustics. Until quite recently all such theories neglected any thermodynamic contribution [13, 14, 15, 16, and 17]. Based on historical perspective, we might reasonably inquire about the potential importance of thermodynamic considerations for electroacoustics. Shilov and others [18] have addressed this query and revealed a new interesting feature of the electroacoustic effect.

Table 2 lists important steps in the development of our understanding of sound. From the very beginning sound was considered as a rather simple example that allowed development of a general theory of "wave" phenomena. Then, later, the new understanding achieved for sound was extended to other wave phenomena, such as light. Tyndall, for example, used reference to sound to explain the wave nature of the light [19, 20]. Newton's Corpuscular Theory of Light was first opposed both by the celebrated astronomer Huygens and the, no less celebrated, mathematician Euler. They each held that light, like *sound*, was a product of wave-motion. In the case of *sound*, the velocity depends upon the relation of elasticity to density in the body that transmits the sound. The greater the elasticity the greater is the velocity, and the less the density the greater is the velocity. To account for the enormous velocity of propagation in the case of light, the substance that transmits it is assumed to have both extreme elasticity and extreme density.

This dominance of sound over light as examples of the wave phenomena continued even with Lord Rayleigh, who developed his theory of scattering mostly for sound and paid much less attention to light [6, 21-24]. At the end of the 19th century sound and light parted because further investigation was directed more on the physical roots of each phenomenon instead of on their common wave nature.

The history of light and sound in Colloid Science is very different. Light has been an important tool since the first microscopic observations of Brownian motion and the first electrophoretic measurements. It became even more important in middle of the 20th century through the use of light scattering for the determination of particle size.

In contrast, sound remained unknown in Colloid Science, despite a considerable amount of work in the field of Acoustics using fluids that were essentially of a colloidal nature. The goal of these studies was to learn more about Acoustics, but not about colloids. This is the spirit in which the ECAH theory (Epstein-

Carhart-Allegra-Hawley [25, 26]) for ultrasound propagation through dilute colloids was developed.

Although Acoustics was not used specifically for colloids, it was a powerful tool for other purposes. For instance, it was used to learn more about the structure of pure liquids and the nature of chemical reactions in liquids. These studies are associated with the name of Prof. Eigen, who received a Nobel Prize in 1968 [27-29].

It is curious that the penetration of ultrasound into Colloid Science began with electroacoustics, which is more complex than traditional acoustics. An Electroacoustic effect was predicted for ions by Debye in 1933 [30], and later extended to colloids by Hermans and, independently, Rutgers in 1938 [31]. The early experimental electroacoustic work is associated with Yeager and Zana, who conducted many experiments in the 1950's and 60's with various co-authors [32-35]. Later this work was continued by Marlow, O'Brien, Ohshima, Shilov, and the authors of this book [13, 15, 16, 17, 18, 36, and 37]. As a result, there are now several commercially available electroacoustic instruments for characterizing ζ-potential.

Acoustics only attained some recognition in the field of colloid science very recently. It was first suggested as a particle sizing tool by Cushman and others [77] in 1973, and later refined by Uusitalo and others [77], and for large particles by Riebel [38]. Development as a commercial instrument having the capability to

102

Table 2 Key Steps in understanding Sound related to Colloids.

Year	Author	Topic
1687	Newton [6]	Sound speed in fluid, theory, erroneous isothermal assumption
Early 1800's	Laplace [6]	Sound speed in fluid, theory, adiabatic assumption
1820	Poisson [51,52]	Scattering by atmosphere arbitrary disturbance, first successful theory
1808	Poisson [51,52]	Reflection from rigid plane, general problem
1845-1851	Stokes [9,10]	Sound attenuation in fluid, theory, viscous losses
1842	Doppler [53]	Alternation of pitch by relative motion
1868	Kirchhoff [7,8]	Sound attenuation in fluid, theory, thermal losses
1866	Maxwell [54]	Kinetic theory of viscosity
1870-80	Henry, Tyndall, Reynolds [19,20,55,56]	First application to colloids - sound propagation in fog
1871	Rayleigh [21,22]	Light scattering theory
1875-80	Rayleigh [6,23,24]	Diffraction and scattering of sound, Fresnel zones in Acoustics
1878	Rayleigh [6]	Theory of Sound, Vol. II
1910	Sewell [12]	Viscous attenuation in colloids, theory
1933	Debye [30]	Electroacoustic effect, introduction for ions
1936	Morse [3]	Scattering theory for arbitrary wavelength-size ratio
1938	Hermans [31]	Electroacoustic effect, introduced for colloids
1944	Foldy [57,58]	Acoustic theory for bubbles
1948	Isakovich [11]	Thermal attenuation in colloids, theory
1946	Pellam, Galt [59]	Pulse technique
1947	Bugosh, Yaeger [32]	Electroacoustic theory for electrolytes
1951-3	Yeager, Hovorka, Derouet, Denizot [33-35,60]	First electroacoustic measurements

Table 2 Key Steps in understanding sound related to Colloids. *(Continued)*

Year	Author	Topic
1951-2	Enderby, Booth [61-62]	First electroacoustic theory for colloids
1953	Epstein and Carhart [25]	General theory of sound attenuation in dilute colloids
1958-9	Happel, Kuwabara [63-65]	Hydrodynamic cell models
1962	Andreae et al [66,67]	Multiple frequencies attenuation measurement
1967	Eigen et al [27-28]	Nobel price, acoustics for chemical reactions in liquids
1972	Allegra, Hawley [26]	ECAH theory for dilute colloids
1973	Cushman [77]	First patent for acoustic particle sizing
1974	Levine, Neale [68]	Electrokinetic cell model
1978	Beck [36]	Measurement of ζ-potential by ultrasonic waves
1981	Shilov, Zharkikh [69]	Corrected electrokinetic cell model
1983	Marlow, Fairhurst, Pendse [36]	First electroacoustic theory for concentrates
1983	Uusitalo [76]	Mean particle size from acoustics, patent
1983	Oja, Peterson, Cannon [70]	ESA electroacoustic effect
1988	Harker, Temple [71]	Coupled phase model for acoustics of concentrates
1987	Riebel [38]	Particle size distribution, patent for the large particles
1988-9	O'Brien [13,15]	Electroacoustic theory, particle size and ζ-potential from electroacoustics
1990	Anson, Chivers [47]	Materials database
1999	Shilov and others [16,17]	Electroacoustic theory for CVI in concentrates
1990 to present	McClements, Povey [43,44,45]	Acoustics for emulsions
1996 to present	A.Dukhin, P.Goetz [39,72,73]	Combining together acoustics and electroacoustics for Particle sizing, Rheology and Electrokinetics

measure a wide particle size range, was begun by Goetz, A.Dukhin, and Pendse in the 90's [39, 40, 41,42]. At the same time a group of British scientists, including McClements, Povey, and others [43, 44, 45, 46, 47, and 48], actively promoted it, especially for emulsions. There are now four commercially available acoustic spectrometers, manufactured by Malvern, Sympatec, Matec, and Dispersion Technology.

To conclude this short historical review we would like to mention a development that we consider of great importance for the future, namely the combination of both acoustic and electroacoustic spectroscopies. The synergism of this combination is described in papers and patents by A.Dukhin and P.Goetz [17, 37, 40, 41, 42, 49, 50, 72, and 73].

Advantages of ultrasound over traditional characterization techniques

There is one major advantage of ultrasound-based techniques compared to traditional characterization methods. Ultrasound can propagate through concentrated suspensions and consequently allows one to characterize concentrated dispersions as is, without any dilution. This feature of ultrasound is applicable to both particle size and ζ-potential measurement. Dilution required by traditional techniques can destroy aggregates or flocs and the corresponding measured particle size distribution for that dilute system would not be correct for the original concentrated sample.

Elimination of dilution is especially critical for ζ-potential characterization, because this parameter is a property of both the particle and the surrounding liquid; dilution changes the suspension medium and, as a result, ζ-potential.

The many advantages of ultrasound for characterizing particle size are summarized in Table 3.

Acoustic methods are very robust and precise [72,73]. They are much less sensitive to contamination compared to traditional light-based techniques, because the high concentration of particles in a fresh sample dominates any small residue from the previous sample. It is a relatively fast technique. Normally a single particle size measurement can be completed in a few minutes. This feature, together with the ability to measure flowing systems, makes acoustic attenuation very attractive for monitoring particle size on-line.

There are several advantages of ultrasound over light based instruments because of the longer wavelength used. The wavelength of ultrasound in water, at the highest frequency typically used (100 MHz), is about 15 microns, and it increases even further to 1.5 millimeters at the lowest frequency (1 MHz). In contrast, light based instruments typically use wavelengths on the order of 0.5 microns. If the particles are small compared to the wavelength we say that this satisfies the Rayleigh long wavelength requirement. It is known that particle sizing in this long wavelength range is more desirable than in the intermediate or short wavelength range, because of lower sensitivity to shape factors and also a simpler theoretical interpretation. As a result, using the longer wavelengths available through acoustics allows us to characterize a much wider range of particle size, while still meeting this long wavelength requirement.

Table 3 Features and benefits of acoustics over traditional particle sizing techniques.

Feature	Benefit
No dilution required.	Less sensitive to contamination
No calibration with the known particle size	More accurate
Particle size range from 5 nm to 1000 microns with the same sensor.	Simpler hardware, more cost effective
Simple decoupling of sound adsorption and sound scattering	Simplifies theory
Possible to eliminate multiple scattering even at high volume fractions up to 50%vol	Simplifies theory for large particle size
Existing theory for ultrasound absorption in concentrates with particles interaction	Possible to treat small particles in concentrates
Data available over wide range of wavelength	Allows use of simplified theory and reduces particle shape effects
Innate weight basis, lower power of the particle size dependence	Better for polydisperse systems
Particle sizing in dispersions with several dispersed phases (mixed dispersions)	Real world, practical systems
Particle sizing in structured dispersions.	

Nature provided one more significant advantage of ultrasound over light, and that is related to the wavelength dependence. As the wave travels through the colloid, it is known that the extinction of both ultrasound and light occurs due to the combined effects of both scattering and absorption [3, 74]. Since most light scattering experiments are performed at a single wavelength it is not possible to experimentally separate these two contributions to the total extinction. In fact, more often than not, the absorption of light is simply neglected in most light scattering experiments, and this can lead to errors.

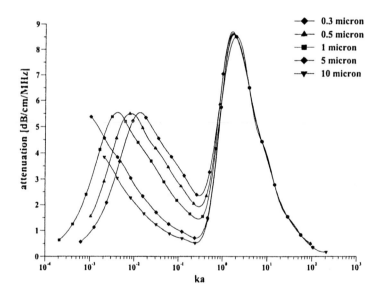

Figure 1. Scattering attenuation and viscous absorption of ultrasound.

In the case of ultrasound, the absorption and scattering are distinctively separated on the wavelength scale. Figure 1 illustrates the dependence of ultrasound attenuation as a function of relative wavelength ka defined by:

$$ka = \frac{2\pi a}{\lambda}$$

where a is the particle radius and λ is the wavelength of ultrasound.

It is seen that attenuation curve has two prominent ranges. The low frequency region corresponds to absorption; the higher frequency region corresponds to scattering. It is obvious from inspection of Figure 1 that it is a simple matter to separate both contributions because there is very little, indeed almost negligible, overlap.

This peculiar aspect of ultrasound frequency dependence allows one to simplify the theory tremendously. Indeed, in the wide majority of cases absorption and scattering can be considered separately. This simplification is valid except for very high volume fractions and for some special non-aqueous systems with soft particles [75].

Electroacoustics is a relatively new technique compared to acoustics. In principle it can provide information for both particle sizing and ζ-potential characterization. However, we believe that acoustics is much more suited to particle sizing than electroacoustics. For this reason, justified in the book [1], we consider electroacoustics as primarily a technique for characterizing only the electric surface properties like ζ-potential. In this sense electroacoustics competes with microelectrophoresis and other traditional electrokinetic methods. However, electroacoustics has many advantages over traditional electrokinetic methods that can be summarized as:

- no dilution required, volume fraction up to 50%vl;
- less sensitive to contamination, easier to clean;
- higher precision ($\pm$ 0.1 mV);
- low surface charges (down to 0.1 mV);
- electrosmotic flow is not important;
- convection is not important;
- faster.

In addition, electroacoustic probes can be used for various titration experiments, as it will be shown in the book [1].

The third field where ultrasound competes with traditional colloid characterization is the field of rheology. This is relatively new area of ultrasound application. We can count two obvious advantages of ultrasound over traditional rheometers. First, ultrasound measurements are non-destructive and allow us to obtain information about the high frequency rheological properties while keeping the sample intact. The second advantage is related to the ability to characterize volume viscosity in addition to shear viscosity. This was already known to Stokes 150 years ago [9]. Volume viscosity is a more sensitive probe of any structural features in a system but it is impossible to measure using shear-based techniques. Ultrasound attenuation is the only known technique able to characterize this important rheological parameter.

In conclusion, we think that the combination of acoustics and electroacoustics enhances each of them [39]. In addition, there is certain overlap in their nature, that offers a way to create various consistency tests to verify the reliability of the data.

REFERENCES

1. Dukhin, A.S. and Goetz, P.J. "Ultrasound for Characterizing Colloids. Particle sizing, Zeta Potential, Rheology", Elsevier, (2002)

2. Lyklema, J. "Fundamentals of Interface and Colloid Science", Volumes 1, Academic Press, (1993)

3. Hunter, R.J. "Foundations of Colloid Science", Oxford University Press, Oxford, (1989)

4. Morse, P. and Ingard, U. "Theoretical Acoustics", McGraw-Hill, NY, (1968)

5. Kinsler, L., Frey, A., Coppens, A. and Sanders, J. "Fundamentals of Acoustics", J. Wiley & Sons, NY, (2000)

6. Rayleigh, L. "The Theory of Sound", Vol.2, Macmillan and Co., NY, second edition 1896, first edition (1878).

7. Kirchhoff, Pogg. Ann., vol. CXXXIV, p.177, (1868)

8. Kirchhoff, "Vorlesungen uber Mathematische Physik", (1876)

9. Stokes, "On a difficulty in the Theory of Sound", Phil. Mag., Nov. (1848)

10. Stokes, "Dynamic Theory of Diffraction", Camb. Phil. Trans., IX, (1849)

11. Isakovich, M.A. Zh. Experimental and Theoretical Physics, 18, 907 (1948)

12. Sewell, C.T.J., "The extinction of sound in a viscous atmosphere by small obstacles of cylindrical and spherical form", PhilTrans. Roy. Soc., London, 210, 239-270 (1910)

13. O'Brien, R.W. "Electro-acoustic Effects in a dilute Suspension of Spherical Particles", J. Fluid Mech., 190, 71-86 (1988)

14. Hunter, R.J. "Review. Recent developments in the electroacoustic characterization of colloidal suspensions and emulsions", Colloids and Surfaces, 141, 37-65 (1998)

15. O'Brien, R.W. "Determination of Particle Size and Electric Charge", US Patent 5,059,909, Oct.22, (1991)

16. Dukhin, A.S., Shilov, V.N., Ohshima, H., Goetz, P.J "Electroacoustics Phenomena in Concentrated Dispersions. New Theory and CVI Experiment", Langmuir, 15, 20, 6692-6706, (1999)

17. Dukhin, A.S., Shilov, V.N, Ohshima, H., Goetz, P.J "Electroacoustics Phenomena in Concentrated Dispersions. Effect of the Surface Conductivity", Langmuir, 16, 2615-2620 (2000)

18. Shilov, V.N. and Dukhin A.S. "Sound-induced thermophoresis and thermodiffusion in electric double layer of disperse particles and electroacoustics of concentrated colloids." Langmuir, submitted.

19. Tyndall, J. "Light and Electricity", D. Appleton and Com., NY (1873).

20. Tyndall, J. "Sound", Phil. Trans., 3rd addition, (1874)

21. Rayleigh, J.W. "On the Light from the Sky", Phil. Mag., (1871)

22. Rayleigh, J.W. "On the scattering of Light by small particles", Phil. Mag., (1871)

23. Rayleigh, J.W. "Acoustical Observations", Phil. Mag., vol IX, p. 281, (1880)

24. Rayleigh, J.W. "On the Application of the Principle of Reciprocity to Acoustics", Royal Society Proceedings, vol XXV, p. 118, (1876)

25. Epstein, P.S. and Carhart R.R., "The Absorption of Sound in Suspensions and Emulsions", J. of Acoust. Soc. Amer., 25, 3, 553-565 (1953)

26. Allegra, J.R. and Hawley, S.A. "Attenuation of Sound in Suspensions and Emulsions: Theory and Experiments", J. Acoust. Soc. Amer., 51, 1545-1564 (1972)

27. Eigen., "Determination of general and specific ionic interactions in solution", Faraday Soc. Discussions, , No.24, p.25 (1957)

28. Eigen, M. and deMaeyer, L. in "Techniques of Organic Chemistry", (ed. Weissberger) Vol. VIII Part 2, Wiley (1963)

29. DeMaeyer, L., Eigen, M., and Suarez, J. "Dielectric Dispersion and Chemical Relaxation", J. Of the Amer. Chem. Soc., 90, 12, 3157-3161(1968)

30. Debye, P. J. Chem. Phys., 1, 13 (1933)

31. Hermans, J. Philos. Mag., 25, 426 (1938)

32. Bugosh, J., Yeager, E. and Hovorka, F. J. Chem. Phys. 15, 592 (1947)

33. Yeager, E. and Hovorka, F. J. Acoust. Soc. Amer., 25, 443 (1953)

34. Yeager, E., Dietrick, H. and Hovorka, F. J. Acoust. Soc. Amer., 25, 456 (1953)

35. Zana, R. and Yeager, E. J. Phys. Chem, 71, 4241 (1967)

36. Beck et al., "Measuring Zeta Potential by Ultrasonic Waves", Tappi, vol.61, 63-65, (1978)

37. Dukhin, A.S., Shilov, V.N. and Borkovskaya. Yu. "Dynamic Electrophoretic Mobility in Concentrated Dispersed Systems. Cell Model.", Langmuir, 15, 10, 3452-3457 (1999)

38. Riebel, U. et al. "The Fundamentals of Particle Size Analysis by Means of Ultrasonic Spectrometry" Part. Part. Syst. Charact., vol.6, pp.135-143, (1989)

39. Dukhin, A.S. and Goetz, P.J. "Acoustic and Electroacoustic Spectroscopy", Langmuir, 12, 19, 4336-4344 (1996)

40. Dukhin, A.S. and Goetz, P.J. "Characterization of aggregation phenomena by means of acoustic and electroacoustic spectroscopy", Colloids and Surfaces, 144, 49-58 (1998)

41. Dukhin, A.S. and Goetz, P.J. "Method and device for characterizing particle size distribution and zeta potential in concentrated system by means of Acoustic and Electroacoustic Spectroscopy", patent USA, 09/108,072, (2000)

42. Dukhin, A.S. and Goetz. P.J. "Method and device for Determining Particle Size Distribution and Zeta Potential in Concentrated Dispersions", patent USA, pending

43. Povey, M. "The Application of Acoustics to the Characterization of Particulate Suspensions", in Ultrasonic and Dielectric Characterization Techniques for Suspended Particulates, ed. V. Hackley and J. Texter, Am. Ceramic Soc., Ohio, (1998)

44. McClements, J.D. "Ultrasonic Determination of Depletion Flocculation in Oil-in-Water Emulsions Containing a Non-Ionic Surfactant", Colloids and Surfaces, 90, 25-35 (1994)

45. McClements, D.J. "Comparison of Multiple Scattering Theories with Experimental Measurements in Emulsions" The Journal of the Acoustical Society of America, vol.91, 2, pp. 849-854, February (1992)

46. Holmes, A.K., Challis, R.E. and Wedlock, D.J. "A Wide-Bandwidth Study of Ultrasound Velocity and Attenuation in Suspensions: Comparison of Theory with Experimental Measurements", J. Colloid and Interface Sci., 156, 261-269 (1993)

47. Anson, L.W. and Chivers, R.C. "Thermal effects in the attenuation of ultrasound in dilute suspensions for low values of acoustic radius", Ultrasonic, 28, 16-25 (1990)

48. Holmes, A.K., Challis, R.E. and Wedlock, D.J. "A Wide-Bandwidth Ultrasonic Study of Suspensions: The Variation of Velocity and Attenuation with Particle Size", J. Colloid and Interface Sci., 168, 339-348 (1994)

49. Dukhin, A.S. and Goetz, P.J. "Acoustic Spectroscopy for Concentrated Polydisperse Colloids with High Density Contrast", Langmuir, 12, [21] 4987-4997 (1996)

50. Wines, T.H., Dukhin A.S. and Somasundaran, P. "Acoustic spectroscopy for characterizing heptane/water/AOT reverse microemulsion", JCIS, 216, 303-308 (1999)

51. Poisson, "Sur l'integration de quelques equations lineaires aux differnces prtielles, et particulierement de l'equation generalie

du mouvement des fluides elastiques", Mem., de l'Institut, t.III, p.121, (1820)

52. Poisson, Journal de l'ecole polytechnique, t.VII, (1808)

53. Doppler, "Theorie des farbigen Lichtes der Doppelsterne", Prag, (1842)

54. Maxwell, "On the Viscosity or Internal Friction of Air and other Gases", Phil. Trans. vol 156, p.249, (1866)

55. Henry, Report of the Lighthouse Board of the United States for the year 1874.

56. Reynolds, O. Proceedings of the Royal Society, vol. XXII, p.531, (1874)

57. Foldy, L.L "Propagation of sound through a liquid containing bubbles", OSRD Report No.6.1-sr1130-1378, (1944)

58. Carnstein, E.L. and Foldy, L.L "Propagation of sound through a liquid containing bubbles", J. of Acoustic Society of America, 19, 3, 481- 499 (1947)

59. Pellam, J.R. and Galt, J.K. "Ultrasonic propagation in liquids: Application of pulse technique to velocity and absorption measurement at 15 Megacycles", J. of Chemical Physics, 14, 10 , 608-613 (1946)

60. Derouet, B. and Denizot, F. C. R. Acad. Sci., Paris, 233,368 (1951)

61. Booth, F. and Enderby, J. "On Electrical Effects due to Sound Waves in Colloidal Suspensions", Proc. of Amer. Phys. Soc., 208A, 32 (1952)

62. Enderby, J.A. "On Electrical Effects Due to Sound Waves in Colloidal Suspensions", Proc. Roy. Soc., London, A207, 329-342 (1951)

63. Happel J. and Brenner, H, "Low Reynolds Number Hydrodynamics", Martinus Nijhoff Publishers, Dordrecht, The Netherlands, (1973)

64. Happel J., "Viscous flow in multiparticle systems: Slow motion of fluids relative to beds of spherical particles", AICHE J., 4, 197-201 (1958)

65. Kuwabara, S. "The forces experienced by randomly distributed parallel circular cylinders or spheres in a viscous flow at small Reynolds numbers", J. Phys. Soc. Japan, 14, 527-532 (1959)

66. Andreae, J. and Joyce, P. "30 to 230 Megacycle Pulse Technique for Ultrasonic Absorption Measurements in Liquids", Brit. J. Appl. Phys., v.13, p.462-467 (1962)

67. Andreae, J., Bass, R., Heasell, E., and Lamb, J. "Pulse Technique for Measuring Ultrasonic Absorption in Liquids", Acustica, v.8, p.131-142 (1958)

68. Levine, S. and Neale, G.H. "The Prediction of Electrokinetic Phenomena within Multiparticle Systems.1.Electrophoresis and

Electroosmosis.", J. of Colloid and Interface Sci., 47, 520-532 (1974)

69. Shilov, V.N., Zharkih, N.I. and Borkovskaya, Yu.B. "Theory of Nonequilibrium Electrosurface Phenomena in Concentrated Disperse System.1.Application of Nonequilibrium Thermodynamics to Cell Model.", Colloid J., 43,3, 434-438 (1981)

70. Oja, T., Petersen, G. and Cannon, D. "Measurement of Electric-Kinetic Properties of a Solution", US Patent 4,497,208, (1985)

71. Harker, A.H. and Temple, J.A.G., "Velocity and Attenuation of Ultrasound in Suspensions of Particles in Fluids", J.Phys.D.:Appl.Phys., 21, 1576-1588 (1988)

72. Dukhin, A.S. and Goetz, P.J. "Acoustic and Electroacoustic Spectroscopy for Characterizing Concentrated Dispersions and Emulsions", Adv. In Colloid and Interface Sci., 92, 73-132 (2001)

73. Dukhin, A.S. and Goetz. P.J. "New Developments in Acoustic and Electroacoustic Spectroscopy for Characterizing Concentrated Dispersions", Colloids and Surfaces, 192, 267-306 (2001)

74. Bohren, C. and Huffman, D. "Absorption and Scattering of Light by Small Particles", J. Wiley & Sons, (1983)

75. Babick, F., Hinze, F. and Ripperger, S. "Dependence of Ultrasonic Attenuation on the Material Properties", Colloids and Surfaces, 172, 33-46 (2000)

76. Uusitalo, S.J., von Alfthan, G.C., Andersson, T.S., Paukku, V.A., Kahara, L.S. and Kiuru, E.S. "Method and apparatus for determination of the average particle size in slurry", US Patent 4,412,451 (1983)

77. Cushman and oth. US Patent 3,779,070 (1973)

Chapter 9

Assessment of Electrostatic Interactions in Dense Charged Colloidal Suspensions Using Frequency Domain Photon Migration

Yingqing Huang, Zhigang Sun, and Eva M. Sevick-Muraca*

Department of Chemical Engineering, Texas A&M University, College Station, TX 77843-3122
*Corresponding author: phone: 979–458–3206; fax: 979–845–6446; email: sevick@che.tamu.edu

The electrostatic repulsive force between particles impacts the colloidal structure, which mediates the bulk properties of the suspensions and dramatically hinders visible light scattering. In this investigation, we explored the impact of effective surface charge and ionic strength on the multiple scattering measured by frequency domain photon migration (FDPM). First, we employed FDPM to measure the isotropic scattering coefficients of dialyzed polystyrene (PS) latex suspensions at 687 and 828 nm as a function of ionic strengths at 65, 25 and 5 mM NaCl equivalents. Measured isotropic scattering coefficients decreased with decreasing ionic strength of the suspensions, suggesting that changes in electrostatic interactions could be evaluated from ensemble measurements of multiply scattered light. At each ionic strength, the isotropic scattering coefficients at varying colloidal volume fractions were regressed to scattering theory which incorporated the mean spherical approximation (MSA) with hard sphere Yukawa (HSY) interaction and Primary model (PM) interaction for monodisperse suspensions in order to yield an effective surface charge given the particle size and ionic strength. The estimates of surface charges obtained from scattering data at 687 and 828 nm were consistently similar but varied with ionic strength. We further investigated the impact of effective surface charge on the multiple scattering through modifying the surface by Rhodamine 6G (R6G) adsorption. Measurements were conducted on dense suspensions (volume fraction = 0.186) at an ionic strength of 5 mM NaCl equivalents with varying amount of positively charged R6G adsorbed on the particle surface. FDPM detected the increase in isotropic scattering coefficient due to decreased electrostatic interaction as a result of R6G adsorption. The corresponding effective surface charge fitted using hard sphere Yukawa interaction model and mean spherical approximation decreased as R6G concentration increased. FDPM can be a potential tool for assessing electrostatic interaction in charged dense suspensions. This research is supported by National Science Foundation (CTS-9876583).

Introduction

The interactions among colloidal particles determine the local structure, which mediates rheology, colloidal stability and impact light scattering efficiency. The structure of a suspension not only should be taken into account when sizing particles using ensemble light scattering techniques, but can also provide information about particle interactions within an ensemble (1, 2).

Typically, the static structure of complex fluids can be obtained from small angle light, neutron, and X-ray scattering measurements. However, concentrated colloidal suspensions multiply scatter light (3, 4), making small angle light scattering measurements impractical without tedious refractive index matching, which itself may affect the interaction, and therefore the structure. Neutron and X-ray scattering require a nuclear reactor or a synchrotron source, which limit their practical and ubiquitous application in complex fluids. In addition, these techniques are limited to nanometer-sized colloids (< 100 nm).

In industrial processes involving dense dispersions such as emulsions, techniques capable of directly characterizing dense suspensions are required, since dilution alters original solvent conditions, and therefore alters interparticle interaction and structures.

Although a few techniques have recently been developed to characterize electrostatic interactions in dense suspensions, the measurement of parameters that govern electrostatic interactions remains elusive. For example, electroacoustic techniques (5, 6) have been applied in measuring zeta potential as well as particle size distribution of concentrated suspensions. Values representing zeta potentials obtained from electroacoustic sonic amplitude (ESA) measurements or colloidal vibration potential (CVP) measurements differ from those obtained from standard electrophoresis (7, 8). Furthermore, these perturbative techniques require an oscillatory electric force or sonic field, which may disturb the original position correlations among particles and resultantly, the structure of the dispersions.

In our laboratory, we have used frequency domain photon migration techniques for accurate and precise measurements of isotropic scattering coefficients which are sensitive to microstructure (9, 10).

In brief, FDPM involves launching an intensity sinusoidally modulated light wave in a multiply scattering medium through an optical fiber. When the light wave propagates through the media, its intensity attenuated and it is phase shifted relative to the incident light. The "photon density wave" propagating through the medium is collected by another fiber optic located some distance away from the source. By approximating the transportation of light wave as a photon diffusion process and applying diffusion theory, the relative values of average intensity (DC), amplitude (AC) and phase shift (PS) measured as a function of distance away from the source or modulation frequency can be used to independently extract the isotropic scattering coefficient and absorption coefficient. Using FDPM, Sun *et. al.* successfully obtained size information of dense monodisperse, polydisperse, and bidisperse suspensions in which volume exclusion effects dominated (11, 12, 13, 14). FDPM has been also been showed capable of probing the structure of dense suspensions of volume exclusion interaction and electrostatic interactions respectively (15, 16).

Historically, the hard sphere model is widely used to model colloidal particle interactions. Actually, almost all colloidal suspensions are charged owing to charged surface groups and adsorption of ions from the solution.

In this work, we seek to extend time-dependent multiple light scattering techniques to assess the impact of ionic strength and effective surface charge upon the electrostatic interaction.

Background

Interaction models among charged particles

The hard – sphere Yukawa interaction (HSY) model and primary interaction (PM) model are two of the most popularly used models describing electrostatic among charged particles (17).

HSY describes the interaction among a monodisperse charged suspension as hard sphere interaction with a Yukawa tail:

$$u(r) = \begin{cases} \infty & r < \sigma \\ e^2 z_{eff}^2 <\sigma> \exp(-\kappa(r-\sigma))/(4\pi\varepsilon\varepsilon_o r) & r > \sigma \end{cases} \tag{1}$$

where e is electron charge; ε_0 is the electric permittivity of vacuum; and ε is dielectric constant of the suspending medium. The parameter σ is the particle diameter; z_{eff} is the effective particle surface charge; and κ is the inverse Debye screening length. In HSY, a charged particle is surrounded by a cloud of counter ion layer, whose thickness is characterized by Debye screen length, κ^{-1}, which decreases with increasing ionic strength. The charged particles repulsively interact through volume exclusion and double layer overlapping. The HSY model is also usually termed as one component model (OCM) (17).

The primary model describes the interactions among charged colloidal particles as direct Columbic interactions:

$$u_{ij}(r) = \begin{cases} \infty & r \le \sigma_{ij} \\ e^2 z_i z_j /(4\pi\varepsilon\varepsilon_o r) & r > \sigma_{ij} \end{cases} \tag{2}$$

Where $\sum_i \rho_i z_i = 0$, and ρ_i is the number density of the component i in the colloidal mixture. In the primary model, counter ions are also considered as components in the dispersions, and presence of the counter ions will not impact the direct interaction among charged particles significantly.

Scattering properties of dense colloidal suspension

For a well-characterized monodisperse colloidal suspension, the isotropic scattering coefficients of the suspension, μ_s, can be predicted by:

$$\mu_s'(\lambda) = \frac{6\varphi}{k^2\sigma^3} \int_0^\pi S(\sigma,\lambda,\theta,\varphi)F(\sigma,\lambda,\theta)sin\theta(1-cos\theta)d\theta \qquad (3)$$

where F is form factor, which addresses the scattering amplitude of light with optical wave length λ at scattering angle θ by a particle of diameter σ. The form factor can be calculated from Mie theory. The structure factor $S(q)$ accounts for the interference effects of scattered light from different particles, at wave vector $q = \frac{4\pi m}{\lambda}sin(\theta/2)$ and contains the information of particle position correlation, and therefore of structure information. Here, m is the refractive index of the medium. In diluted suspensions (volume fractions < 1%), particles are randomly oriented and scatter independently. In this case, the interference of scattered light is not significant, and $S(q)$ is close to one. The structure factor $S(q)$ can be predicted with Ornstein-Zernike (O-Z) integral equation employing a first principles model of interaction, such as HSY or PM models, and a approximate closure model relating structure with the interaction, such as mean spherical approximation (MSA), and Percus-Yevick (PY) models (17).

The analytical forms of structure model of monodisperse suspensions are available through integral equation approach using MSA-HSY (18, 19). The analytical solution by Herrera *et al.* (19) is used owing to its simplicity.

Blum and Hoye (20, 21) solved the O-Z equation analytically using MSA associated with PM interaction model. Hiroike (22, 23) reorganized Blum's solution and derived an explicit expression of direct correlation function, whose Fourier transform can be used to directly calculate the partial structure factor $S_{ij}(q)$.

In this study, two analytical solutions of the structure models from the solution of O-Z equation using MSA-HSY (Herrera *et. al.*) and MSA-PM (23) closure and interaction potential models are used to fit the experimental data of isotropic scattering coefficient as a function of ionic strength and volume fraction.

Theory of frequency domain photon migration

FDPM is based on photon diffusion theory which assumes that the transport of multiple scattering light in dense suspension can be approximated by "random walk" of photon, and one can use "frequency domain photon diffusion equation" to describe the optical fluence, Φ, modulated at frequency, ω, at position r in terms of absorption coefficient, μ_a, and isotropic scattering coefficient, μ_s'.

$$D\nabla^2\Phi_{AC}(r,\omega) - [\mu_a - \frac{i\omega}{c}]\Phi_{AC}(r,\omega) = S(r,\omega) \qquad (4)$$

where $D = 1/3(\mu_a + \mu_s{}')$, is called the optical diffusion coefficient. $S(r,\omega)$ is the source term. In infinite media, the decay of photon density wave can be obtained from the solution of above frequency domain diffusion equation to obtain the photon density $U(r,t) = \Phi/c$, at position r and time t as:

$$U = \frac{1}{4\pi Dcr}(S_{DC}\exp(-\mu_{eff}r) + S_{AC}\exp(-i\mu_{eff}r - i\omega t - \phi_o)) \qquad (5)$$

where $\mu_{eff} = (\mu_a/D)^{1/2}$, and c is the the speed of light. S_{AC}, S_{DC}, and ϕ_o are the average intensity, amplitude and initial phase associated with point optical source.

By applying analytical solution of photon diffusion equation of eq. 4, and taking measurement at multiple detecting positions, we can avoid source information and subsequently source calibration.

$$\ln(\frac{rDC(r)}{r_0 DC(r_0)}) = -(r - r_0)[3\mu_a(\mu_a + \mu_s{}')]^{1/2} \qquad (6)$$

$$\ln(\frac{rAC(r)}{r_0 AC(r_0)}) = -(r - r_0)\sqrt{\frac{3}{2}\mu_a(\mu_a + \mu_s{}')}(\sqrt{1 + (\frac{\omega}{v\mu_a})^2} + 1)^{1/2} \qquad (7)$$

$$PS(r) - PS(r_0) = (r - r_0)\sqrt{\frac{3}{2}\mu_a(\mu_a + \mu_s{}')}(\sqrt{1 + (\frac{\omega}{v\mu_a})^2} - 1)^{1/2} \qquad (8)$$

The above three equations, which relate average intensity DC, amplitude AC, and phase shift PS at source-detector distances r and r_o with the optical properties μ_a and $\mu_s{}'$, are obtained from eq. 5. By measuring AC, DC, and PS at multiple source detector distances, we can extract the optical properties of μ_a and $\mu_s{}'$ independently through standard parameter estimation approach (9).

The typical FDPM setup used in measuring isotropic scattering coefficients of a colloidal system has been detailed elsewhere (24). Herein we employ measurements of relative DC and relative phase at optimal modulating frequencies ranging from 50 MHz to 100 MHz to determine accurate values of isotropic scattering coefficients.

Materials and Experimental Considerations

Samples and sample characterization

Polystyrene samples were obtained from Dow Chemicals (Midland, MI). To remove extra ions and the surfactant used to stabilize the suspensions, the polystyrene lattices were first dialyzed using membrane tube (Spectra/Pro: MWCO 6-8,000, Spectrum laboratories, Inc., Silver Spring, MD) in deionized water (W-20, Fisher) until the conductivity of the equilibrium dialyzing water was less than 6 ppm NaCl equivalents, as measured using a titration controller (Accumet Model 150, Fisher). The volume fractions of stock dialyzed polystyrene lattices were measured by an evaporation method, which involves weighting samples using a 1/10,000g resolution balance (Denver Instrument M-220D, Fisher) before and after samples were dried in an oven (Isotemp Model 280A, Fisher) for 8 hours at 95°C. Weight loss due to water evaporation was used to calculated volume fraction. The dialyzed polystyrene suspensions were then diluted to desired volume fraction using final dialyzing water. 2 M sodium chloride solution (S1240, Spectrum Chemical Mfg. Corp., Gardena, CA) was then added to adjust the dispersion to desired ionic strengths. Volume fractions of samples were recalculated after the addition of NaCl solution.

The mean size (143 ± 2 nm) was measured using dynamic light scattering (Zetasizer 3000HS, Malvern Instruments Inc., Worcestershire, UK). The TEM (Zeiss 10C) images were analyzed using Image Pro software (Image Pro 3.0, MediaCybernetics Inc.) to obtain the standard size deviation of 22 ± 4 nm. The effective surface of about 1,000 electron charge is approximated by $z = \pi\varepsilon\sigma(2+\kappa\sigma)\zeta$ from electrophoretically measured zeta potential, ζ, (Malvern Zetasizer 3000HS) (1).

Rhodamine 6G adsorption on polystyrene

In this study, we adjusted the effective surface charge of PS latex by adsorbing positively charged Rhodamine 6G (R6G) (20, 21). R6G (R- 4127, Sigma, St. Louis, MO) molecules can form a positively charged surfactant and negatively charged counter ion once ionized in water. R6G adsorption on the surface of negatively charged polystyrene surface upon two main mechanisms: (1) ion condensation owing to electrostatic interaction, which may neutralize negatively charged group on PS surface and (2) hydrophobic interaction owing to hydrophobic groups in R6G molecules and on polystyrene surface (25,26).

R6G strongly absorbs light at the wavelength of 529 nm and fluoresces at 566 nm with high quantum efficiency. This enables determination of R6G concentrations using fluorescence measurements. R6G also has minimal light absorption and consequently minimal fluorescence in near infrared region, enabling the evaluation of isotopic scattering of colloids at 687 nm using FDPM measurement without the interference from R6G absorbance or fluorescence.

Results and Discussions

Change of μ_s' with Ionic Strength

Figure 1 shows the isotropic scattering coefficients [cm^{-1}] at wavelengths of 687 and 828 nm as functions of volume fraction at the ionic strengths of 60, 25, and 5 mM NaCl equivalents. The symbols denote FDPM measured values, which evidently decreases as the ionic strength decreases at both wavelengths, suggesting FDPM successfully captured the structure charge owing to changed ionic strength. Isotropic scattering coefficients as a function of volume fraction were fitted using equations (3) with Herrera's (19) MSA-HSY structure model for calculating obtain effective surface, z_{eff}, as the only fitting parameter. Herein, a least squares method is used and iteration converges when the step size of surface charge was less than 1 electron charge, and the relative decrease in the merit function $\chi^2 = \sum (\mu_{s,expt}' - \mu_{s,theory}')^2$, was less than 0.1.

The model calculation (curves) with MSA-HSY using fitted z_{eff} match the FDPM experimental data (symbols) well, suggesting first principles model can be used to describe the multiple scattering from a charged dense suspension. The fitted effective surface charges decrease with decreasing ionic strength, from around 1000 electron charges at 60 mM NaCl equivalents, to 160 at 1 mM NaCl equivalents. At the same ionic strength, the effective charges are similar for data evaluated at the two different wavelengths as shown in Table I. At 25 and 60 mM NaCl equivalents, the fitted surface charge is similar to the value estimated from the zeta potential measurement.

MSA-PM model (22) was also used to regress FDPM experimental data. The regressed effective surface charges are shown in the Table I. Using one effective surface charge at each ionic strength for all volume fractions, MSA-PM model over predicted the scattering at volume fractions higher than 15%. The fitted surface charges using MSA-PM are 1~2 orders of magnitude less than those fitted using MSA-HSY, and generally increase as the ionic strength decreases. The fitted surface charges using both MSA models at 828 nm are greater than those at 687 nm in most cases, suggesting both models captured limited physics.

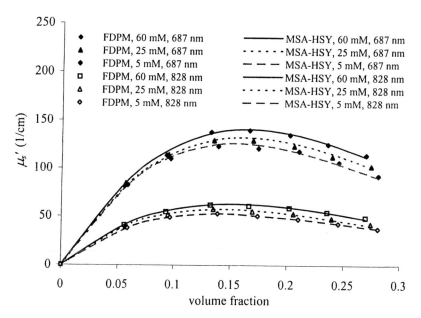

Figure 1: Isotropic scattering coefficients versus volume fraction as a function of wavelength and ionic strength in NaCl equivalents. Symbols denote FDPM measured results, and lines denote the prediction using MSA-HSY model with regressed z_{eff} (Reproduced from reference 27. Copyright 2002 American Chemical Society.)

Table I: Fitted effective average charge of MSA models

Ionic strength NaCl equiv. (mM)		120 mM	60 mM	25 mM	5 mM	1 mM
Effective Charge, e MSA-HSY	687 nm	1.9	1240	836	404	137
	828nm	0	1005	1137	489	184
Effective Charge, e MSA-PM	687 nm	0	7	12	15	15
	828 nm	0	3	9	12	13

We further verified MSA-HSY model by modifying the surface charge using R6G adsorption.

Since the model predictions using MSA-HSY fitted z_{eff} match FDPM experimental data better than the MSA-PM model, and since the fitted effective surface charges are closer to the values obtained from the electrophoretical approach, we conclude that MSA-HSY model describes the structure of charged colloidal suspension better than MSA-PM does.

Change of μ_s' with Rhodamine 6G adsorption

R6G can adsorb on PS surface by hydrophobic or electrostatic interaction. It is reported that at low R6G surface converges, electrostatic interaction dominates, and at higher R6G surface converges, hydrophobic interaction is overwhelming (18, 19).

Figure 2 plots the change in the isotropic scattering coefficient versus total R6G concentration at constant PS volume fraction of 0.186 and constant ionic strength of 5 mM NaCl equivalents. Based upon our experimental determined adsorption isotherm, more than 99.9% of R6G added to the suspension was adsorbed on the polystyrene surface. The amount of the free R6G in the solvent is negligible in comparison to the amount of adsorbed R6G on PS surface. The isotropic scattering coefficient, μ_s', increases with the increasing total R6G concentration, and does not evidently increase after an inflection points (of total R6G concentration about 0.2 mM) with further addition of R6G. Since the R6G surface coverage on polystyrene, (which is calculated from the R6G molecule size and surface adsorption density from the adsorption isotherm,) was relatively low (<10 % before the inflection point), an increase in effective particle size owing to R6G adsorption would be insufficient to change the scattering cross section of each individual particle. This assures that change in the scattering efficiency of each individual particle owing to R6G adsorption could not be responsible for the increased μ_s'. Since trace amount of added R6G cannot significantly alter the ionic strength, especially before the inflection point where even less amount R6G was added, ionic strength change owing to R6G addition

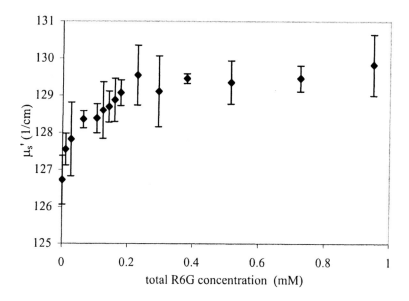

Figure 2: FDPM measured isotropic scattering and absorption coefficients versus total R6G concentration in polystyrene suspension with volume fraction of 0.186 and ionic strength of 5 mM equiv.). (Reproduced from reference 28. Copyright 2003 American Chemical Society.)

can be eliminated as a cause. Therefore, the change in effective surface charge owing to R6G adsorption is the only significant mechanism for the increase in μ_s', and FDPM successfully captured the subtle change in the optical properties owing to the changing effective surface charge as a result of R6G adsorption.

Change in z_{eff} owing to R6G adsorption

Because of possible presence of hydrophobic condensation of R6G on polystyrene surface, one cannot infer the change in surface charge of the polystyrene from the amount of adsorbed R6G by simple neutralization calculation. Herein, MSA-HSY model by Herrera (19) was again used to fit experimentally measured μ_s' to extract the effective surface charge at each addition of R6G. Figure 3 shows the fitted effective surface charge using MSA-

HSY model versus total R6G concentration. Effective surface charges decrease with increasing R6G concentration, and do not change significantly after the inflection point. The typical error in z_{eff} propagated from the uncertainties of FDPM measurement is about 10 electron charges. The existence of the inflection indicates the change in dominant adsorption mechanism from electrostatic condensation to hydrophobic adsorption.

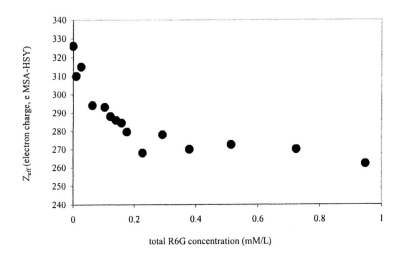

Figure 3: The parameter estimate of z_{eff} versus total R6G concentration in the polystyrene suspension (with volume fraction of 0.186, ionic strength of 5 mM NaCl equiv.). The propagated errors in z_{eff} from the uncertainties of FDPM measurement, about 10 electron charge, is not shown in figure. Herrera's solution to the MSA-HSY was employed (19). (Reproduced from reference 28. Copyright 2003 American Chemical Society.)

In general, FDPM measurement, associated with the MSA-HSY model, displayed sensitivity to changing surface charge owing to the surface adsorption of R6G on polystyrene particle surface. We noted that the surface "charge" from the model prediction may not be the actual physical surface charge, and it depends on the potential model and closure relation to the O-Z equation. However, this suggests that with a better model or with the some kind of charge calibration, FDPM could be used to assess a change in surface charge, or even to detect zeta potentials of dense multiply scattering dispersions.

Conclusion

Electrostatic interaction should be considered when investigating scattering from charged dense suspensions. To charged polystyrene lattices, the increasing repulsive force associated with decreasing ionic strength and/or increasing surface charge causes the polystyrene to be more structured and impacts the scattering at high volume fractions. First principles model of interaction, such as, MSA-HSY can be used to describe multiple light scattering of charged dense suspensions. FDPM measured isotropic scattering coefficients displayed sensitivity to the interaction change owing to the change in ionic strength and z_{eff} and FDPM may provide a potential probe for interaction and structure of dense suspensions.

This work is partially supported by the National Science Foundation. (CTS-9876583)

Reference

1   Hiemenz, P.C.; Rajagopalan, R.   *Principles of Colloid and Surface Chemistry;* 3rd ed.; Marcel & Dekker, 1997; Chapter 11.

2   Hunter, R.J. *Foundations of Colloid Science;* Oxford science, 1987.

3   Fraden S.; Maret G. *Phy. Rev. Lett.* **1990**, *65*, 512-515.

4   Garg, R.; Prud'homme, R.K.; Aksay, I.A.; Liu, F.; Alfano, R.R. *J. Opt. Soc. Am. A.* **1998**, *15*, 932.

5   O'Brien, R.W.; Cannon, D.W.; Rowlands, W.N. *J. Colloid Interface Sci.* **1995**, *173*, 406.

6   Carasso, M.L.; Rowlands, W.R.; Kennedy, R.A. *J.Colloid Interface Sci.* **1995**, *174*, 405.

7   Babchin, A.J.; Chow, R.S.; Sawatzky, R.P. *Adv. Colloid Interface Sci.* **1989**, *30*, 111.

8   Goetz, R.J.; El-Aasser, M.S. *J. Colloid Interface Sci.* **1992**, *150* (2), 436.

9   Fishkin, J.B.; So, P.T.C.; Cerussi, A.E.; Fantini, S.; Franceschini, M.A.; Gratton, E. *Appl. Opt.* **1995**, *34* (7), 1143.

10 Sun, Z.; Huang, Y.; Sevick-Muraca, E.M. *Rev. Sci. Instrum.* **2002**, *73* (2), 383.

11 Sun, Z.; Tomlin, C.D.; Sevick-Muraca, E.M. *AICHE J.* **2001**, *47*, 1487.

12 Sun, Z.; Tomlin, C.D.; Sevick-Muraca, E.M. *J. of Colloid and Interface Sci.* **2002**, *245*, 281.13   Sun, Z.; Tomlin, C.D.; Sevick-Muraca, E.M. *Langmuir* **2001**, *17* (20), 6142.

14 Sun, Z.; Sevick-Muraca, E.M. *Langmuir* **2002**, *18,* 1091-1097.15

    Banerjee, S.; Shinde, R; Sevick-Muraca, E.M. *J. Chem. Phys.* **1999**, *111* (20), 9133.

16 Huang, Y.; Sevick-Muraca, E.M. *J. Colloid Interface Sci. (Accepted)* **2002**.

17 Hansen, J.P.; McDonald I.R. *Theory of Simple Liquids*, 2nd ed.; Academic Press, New York, **1986**.

18 Hayter, J. B.; Penfold, J. *Mol. Phys.* **1981**, *42* (1), 109.

19 Herrera, J.N.; Cummings, P.T.; Ruiz-Estrada, H.R. *Mol. Phys.* **1999**, *5*, 835.

20 Blum L. *Mol. Phys.* **1975**, *30* (5), 1529.

21 Blum, L.; Hoye J. S. *J. of Phys. Chem.* **1977**, *81*(13), 1311.

22 Hiroike, K. *J. of the Physical Society of Japan.* **1969**, *27* (6), 1415.

23 Hiroike, K. *Mol. Phys.* **1977**, *33* (4), 1195.

24 Richter, S.M.; Shinde, R.R.; Balgi, G.V.; Sevick-Muraca, E.M. *Part. Part. Syst. Charact.* **1995**, *15*, 9.

25 Gong, Y.K.; Nakashima, K.; Xu, R. *Langmuir* **2000**, *16*, 8546.

26 Charreyre M.T.; Zhang, P.; Winnik, M.A.; Pichot, C.; Graillat, C. *J. of Colloid and Interface Sci.* **1995**, *170*, 374.

27 Huang, Y.; Sun, Z.; Sevick-Muraca, E.M. *Langmuir* **2002**, *18*, 2048.

28 Huang, Y.; Yuwono, V.; Sevick-Muraca, E.M. *Langmuir*. (Submitted) 2002.

Chapter 10

Novel Fundamental Model for the Prediction of Multiply Scattered Generic Waves in Particulates

Felix Alba

Felix ALBA Consultants, Inc., 5760 South Ridge Creek Road, Murray, UT 84107 (felixalba@scatterer.com)

Fundamental models for accuratetly predicting multiple scattering of electromagnetic and acoustic waves through particulates have been in need for a long time. The lack of these models forces optical size analyzers to only accept highly diluted samples and impairs the performance of acoustic analyzers. After 20 years of R & D this vacuum has been filled with METAMODEL™. This new multiple-scattering model is as fundamental as Mie and ECAH single-scattering models. Its generic validity resides in having modeled the interaction between the fields scattered by every particle, leading to *generic stochastic field equations* where the overlapping of all fields is contemplated regardless of their physical nature (viscous-inertial, thermal diffusion, electromagnetic, elastic, etc.). Its fundamentals are described, and validating data in concentrated dispersions are presented.

Introduction

Electromagnetic waves have been routinely used for the characterization of suspensions and emulsions for more than three decades now. The Mie fundamental mathematical model is the war-horse for predicting the attenuation and intensity pattern of the scattered waves with the proviso that the particle concentration has to be low enough for the phenomenon of multiple scattering to be negligible. *A fundamental mathematical model for high concentrations has been in need for a long time.*

During the last decade, *acoustic waves* have entered the commercial scene as a powerful tool for the characterization of concentrated dispersions. For acoustic waves there exists also a *fundamental* single-scattering mathematical model known as the Epstein-Carhart-Allegra-Hawley (ECAH) model (*1*).

Acoustic and electromagnetic wave phenomena are inherently different. Both obviously exhibit multiple scattering when interacting with particulates. However, in general, multiple scattering of acoustic waves in the ultrasonic range starts being significant at much higher particle concentrations than multiple scattering of light waves. This remarkable feature of ultrasound has been for long time the big promise for accurate characterization of particulates at industrial concentrations.

Having said that, researchers working in *acoustics* know that multiple scattering, at a fixed concentration, will be significant or not depending upon the particular phenomenon dominating the interaction (viscous-inertial, thermal diffusion, etc.), as well as upon the ratio between the particle size and the wavelength. The lower the frequency and the smaller the particle size, the lower the particle concentration beyond which multiple scattering cannot be ignored. The ECAH model can predict very accurately the attenuation and velocity spectra of a submicron polystyrene suspension at a concentration by volume (Cv) of 30% (*1*), but will fail miserably in predicting the same for a submicron silica suspension at Cv=10% (*2*). The dominant phenomenon for the former is thermal diffusion, for the latter is viscous-inertial.

In summary, if the big promise of acoustics it to be fulfilled, a fundamental model for accurate prediction of multiple scattering of *acoustic waves* in particulates is as essential as for *electromagnetic waves*.

Approaches to Mathematical Modeling

There are two basic approaches to modeling the interaction of waves with composites: *microscopic* (fundamental) and *macroscopic* (phenomenological). In the fundamental approach -the Analytical Wave Scattering Theory- the suspension is treated as what it is: a two-phase system, with the element of volume being much smaller than both the smallest particle and the smallest wavelength. This description is the most complex but the most general and accurate (*3*).

In the macroscopic approach -the Radiative (Transport) Theory- the dimension of the volume element is chosen such that it is much smaller than the minimum wavelength transmitted but still big enough to contain a large number of particles. In this fashion, the suspension appears to be homogeneous to the wave. This approach is much simpler, but it delivers not so accurate predictions and it is definitely very limited in scope. To begin with, only situations where the so-called long-wave regime is valid are amenable to this modeling approach (*4*).

The above classification is not sharp: there are some intermediate approaches like the *cell-model, the coupled-phase theory*, and *closed-form approximations* to the fundamental one which avoid the numerical solution of the wave-equations by assuming the wavelength much longer than the size of the particles, the viscous wave penetration depth much smaller than the particle radius, the particle size much larger than the wavelength, etc.

Classical Fundamental Models

Starting with the fundamental work of Lord Rayleigh, single and multiple-scattering models based on the analytical scattering theory evolved during the last century. An incomplete list of main contributors contains Mie, Foldy, Lax, Mal & Bose, Mathur & Yeh, Bringi, Ma, Epstein & Carhart, Waterman & Truell, Twersky, Chow, Fikioris & Waterman, Lloyd & Berry, Allegra & Hawley, Devaney, Tsang & Kong, de Daran, etc. Finally, in 1985, Varadan et al recognized the common mathematical structure behind the description of multiple scattering regardless of the physical nature of the waves, establishing the basic framework needed to develop a generic treatment for electromagnetic, acoustic, and elastic waves (*5*).

METAMODEL™: Fundamental and Generic

Even though there have been published/used several approaches to modeling the complex phenomenon of *multiple scattering*, they are either of a *semi-empirical* nature with severe limitations in size/wavelength range and performance or, albeit having a *fundamental flavor*, have proven not accurate enough to be employed in broad range/application scientific instruments. After 20 years of research and development, this vacuum has been filled with a mathematical model called METAMODEL™ (2).

The METAMODEL™ *multiple-scattering* prediction engine is as fundamental as Mie and ECAH *single-scattering* models. Besides its *fundamental* character, its *generic* validity resides in having described and calculated the detailed interaction between the scattered fields produced by every particle, treating such an interaction in a statistical sense and leading to *generic stochastic field equations* where the overlapping of all scattered fields is equally contemplated regardless of their physical nature (viscous-inertial, thermal diffusion, elastic, electromagnetic, etc.).

METAMODEL™ belongs to the fundamental microscopic category and results from the exhaustive application of the analytical wave scattering theory by integrating viscous-inertial, thermal diffusion, surface tension, and viscoelasticity into the mathematical description of multiple-scattering. In addition, it is generic because it can accurately predict the multiple scattering of *electromagnetic*, *acoustic*, or *elastic* waves interacting with particulates.

Stochastic Field Descriptors

We consider a random spatial distribution of a large number of particles with arbitrary physical attributes (thermodynamical, size, shape, orientation, acoustic, elastic, electromagnetic, transport, etc.) suspended or included in an otherwise continuous medium with an arbitrary concentration. In our stochastic description, we deal with three random fields: the incident (I), the exciting (E), and the scattered (S) fields. For a fixed spatial/attribute distribution of scatterers, these fields obey the vector Hemholtz wave equation:

$$\forall \underline{F} \in \{\underline{I}, \underline{E}, \underline{S}\} \quad (\nabla^2 + k^2)\underline{F}(\underline{r} : \{\underline{r}_1, ..., \underline{r}_i, ..., \underline{r}_N ; \underline{a}_1, ..., \underline{a}_i, ..., \underline{a}_N\}) = 0 \qquad (1)$$

The particular position and attributes of each scatterer are unknown but the ensemble of them can be statistically described through a probability density function $p(\underline{r}_1,...\underline{r}_i,...\underline{r}_N;\underline{a}_1,...,\underline{a}_i,...,\underline{a}_N)$. Foldy introduced in 1945 the concept of a configurational average (6). Being all fields random variables which value depends on the spatial/attribute distribution of the scatterers, the zero-order configurational average $\langle\underline{F}(\underline{r})\rangle$ is simply the mean value for the field (the so-called coherent part of the field):

$$\langle\underline{F}(\underline{r})\rangle = \int...\int\int...\int\underline{F}(\underline{r}:\{\underline{r}_1,...,\underline{r}_i,...,\underline{r}_N;\underline{a}_1,...,\underline{a}_i,...\underline{a}_N\})p(\underline{r}_1,...,\underline{r}_i,...\underline{r}_N;$$

$$\underline{a}_1,...,\underline{a}_i,...,\underline{a}_N)dv_1...dv_i...dv_N d\underline{a}_1...d\underline{a}_i...d\underline{a}_N \tag{2}$$

In a similar way we can define higher order statistics like what Foldy called the first order partial configurational average, i.e., the mean value of the field given that a particle is at a certain location and has a certain set of attributes. The apostrophe in equation 3 means that the averaging does not take place over the known location/attributes with the probability being conditional.

$$\langle\underline{F}(\underline{r}/\underline{r}_i;\underline{a}_i)\rangle = \int...'...\int\int...'...\int\underline{F}(\underline{r}/\underline{r}_i;\underline{a}_i:\{\underline{r}_1,...,\underline{r}_i,...,\underline{r}_N;\underline{a}_1,...,\underline{a}_i,...,\underline{a}_N\})$$

$$p(\underline{r}_1,...,',...,\underline{r}_N;\underline{a}_1,...,',...,\underline{a}_N/\underline{r}_i;\underline{a}_i)dv_1,...'..dv_N d\underline{a}_1...'..d\underline{a}_N \tag{3}$$

Now we observe that the zero-order statistics can be expressed in terms of the first-order statistics by simply performing the missing integral:

$$\langle\underline{F}(\underline{r})\rangle = \int\int\langle\underline{F}(\underline{r}/\underline{r}_i;\underline{a}_i)\rangle p(\underline{r}_i;\underline{a}_i)d\underline{a}_i dv_i \tag{4}$$

Similarly the 1-order statistics can be expressed in terms of the 2-order statistics as follows:

$$\langle\underline{F}(\underline{r}/\underline{r}_i;\underline{a}_i)\rangle = \int\int\langle\underline{F}(\underline{r}/\underline{r}_i,\underline{r}_j;\underline{a}_i,\underline{a}_j)\rangle p(\underline{r}_j;\underline{a}_j/\underline{r}_i;\underline{a}_i)d\underline{a}_j dv_j \tag{5}$$

And, in general, the n-order statistics can be expressed in terms of the n+1-order statistics. Obviously thus we have a recurrent hierarchy of equations which is impractical to pursue all the way down until collapse. Lax (7) proposed the so-called "quasi-crystalline" approximation to break-down the hierarchy as early as possible. By doing so, we have now an integral equation which relates the first order statistics for a given location/attribute with the first order statistics for all other possible locations/attributes:

$$\langle F(\underline{r}/\underline{r}_i;\underline{a}_i)\rangle = \int\limits_{Vi}\int\limits_{D}\langle \underline{F}(\underline{r}/\underline{r}_j;\underline{a}_j)\rangle p(\underline{r}_j;\underline{a}_j/\underline{r}_i;\underline{a}_i)d\underline{a}_j dv_j \qquad (6)$$

Field Equations

For a fixed spatial/attribute distribution of scatterers a field equation can be formulated by simply stating that the exciting field around a particle at a certain location with a certain set of attributes is the summation of the incident field plus all fields scattered by the rest of the particles. There are as many of these equations as possible spatial/attribute distributions of particles exist. We convert this set of deterministic equations into a stochastic one by taking configurational averages on both sides and by letting the number of particles N go to infinity while keeping the concentration constant. The following equation is obtained:

$$\langle \underline{E}(\underline{r}/\underline{r}_i;\underline{a}_i)\rangle = \underline{I}(\underline{r}) + \int\limits_{Vi}\int\limits_{D}\langle \underline{S}(\underline{r}/\underline{r}_j;\underline{a}_j)\rangle n(\underline{r}_j)f_n(\underline{a}_j)g(\underline{r}_j,\underline{a}_j,\underline{r}_i,\underline{a}_i)d\underline{a}_j dv_j \qquad (7)$$

$n(\underline{r}_j)$ is the local particle concentration by number, $f_n(\underline{a}_j)$ is the local density distribution by number of the attributes, and $g(\underline{r}_j,\underline{a}_j,\underline{r}_i,\underline{a}_i)$ is the so-called (well-known in Statistical Mechanics) *pair-correlation function.*

So far we have not talked about the physical nature of the waves as we simply have discussed how to combine all the scattered fields from all particles at a certain observation point. It is in the relation between the field exciting a particle and the field scattered by the same particle where the physical nature of the wave shows its signature. It is well-known that, regardless of the physical character of the wave, the scattered field by a particle in a medium can be expressed as the *"T-transition matrix"* times the exciting field (8). We refer to this matrix as the particle/medium signature. We have here once more a large set of equations -- one for each spatial/attribute distribution of scatterers. Once again, taking configurational averages, we convert the above large set of equations into a single stochastic equation:

$$\langle S(\underline{r}/\underline{r}_j;\underline{a}_j)\rangle = \underline{\underline{T}}(\underline{o},\underline{m},\underline{a}_j)\langle E(\underline{r}/\underline{r}_j;\underline{a}_j)\rangle \tag{8}$$

Replacing now the scattered field in the previous integral equation 7 with the referred signature times the exciting field, and averaging once more with respect to the attributes of the particle at location $\underline{r}_i$, we arrive to an equation relating the exciting field on a particle located at point $\underline{r}_i$ with the exciting fields around the rest of the particles in the ensemble.

$$\langle\langle \underline{E}(\underline{r}/\underline{r}_i)\rangle\rangle = \underline{I}(\underline{r}) + \int_D f_n(\underline{a}_i)\int_{Vi}\int_D \underline{\underline{T}}(\underline{o},\underline{m},\underline{a}_j)\langle\langle \underline{E}(\underline{r}/\underline{r}_j)\rangle\rangle n(\underline{r}_j)f_n(\underline{a}_j)$$

$$g(\underline{r}_j,\underline{a}_j,\underline{r}_i,\underline{a}_i)d\underline{a}_j dv_j d\underline{a}_i \tag{9}$$

This is the basic field integral equation we have to solve.

Solution of the Field Equations

Because all fields verify the vectorial Hemholtz wave equation, they can be expressed as (9):

$$\underline{F}(\underline{r}) = \sum_{n=0}^{\infty}\sum_{m=-n}^{m=n}[f_{nm1}L_{nm}(\underline{r}) + f_{nm2}M_{nm}(\underline{r}) + f_{nm3}N_{nm}(\underline{r})] \tag{10}$$

where L_{nm}, M_{nm}, N_{nm} are eigenvector solutions of the wave equation and $f_{nm1}, f_{nm2}, f_{nm3}$ the linear combination coefficients. The solution finding approach can be summarized as follows:

1. Choose a coordinate system (spherical, cylindrical),
2. Use Bessel or Hankel sets as appropriate with Legendre Polynomials,
3. Replace into field equation 9,
4. Express all fields on a common reference frame,
5. Apply mutual orthogonality of eigenvector sets,
6. Transform volume integrals into surface integrals arriving to an infinite linear homogeneous system,
7. Truncate and find the propagation constant of the composite by looking for a non-trivial solution.

SCATTERER™ : A Long-Needed Tool

In order to conduct research on mathematical modeling over the years I had to devise a flexible software structure which would allow the researcher to easily and efficiently conduct predictions with different mathematical models, parameter sensitivity studies, particle size analysis from data acquired from different spectrometers, etc. As a result, a powerful research and development tool called SCATTERER™ was born (2).

Two numerical engines interact with the researcher through a powerful friendly Windows® interface. The Prediction Engine predicts the composite physical attributes from the physical attributes of the constituent phases and the operating variables. A typical computer experiment would be to predict the attenuation and velocity spectra from the knowledge of both media physical properties and the particle size distribution and concentration.

The *Inversion Engine* handles the inverse problem, i.e. knowing some of the attributes of the composite and constituents; it estimates the values of the missing attributes of the constituents and composite. This engine provides for a generalized direct/invert algorithmic scheme allowing the operator to manipulate the METAMODEL™ mathematical model (and all the others) in any desired direction specifying those parameters that are known/measured as well as those that are not known and are to be estimated. A typical computer experiment would be to determine the particle size distribution (PSD) and concentration from the knowledge of both media physical properties and the measured attenuation spectrum. The SCATTERER™ tool offers three suites: *Acoustics*, *Electromagnetics*, and *Elastodynamics* (2).

Validating METAMODEL™

The most appropriate field to fully validate the METAMODEL™ model is in *Acoustics* because: a) the diversity of physical phenomena occurring when a sound wave interacts with particulates, b) accurate high concentration data have been made available during the last decade due to the recent commercial advent of acoustic spectrometers, and c) it is the successful integration of viscous-inertial, thermal diffusion, surface tension, and viscoelasticity phenomena into the fundamental statistical treatment of multiple scattering what makes METAMODEL™ unique.

Acoustical data in suspensions and emulsions were gathered using the acoustic spectrometer function of a particle size analyzer (ULTRASIZER™) manufactured by Malvern Instruments Ltd. Concentration ladders from low concentrations, where both the standard ECAH single-scattering model and the METAMODEL™ mathematical model perfectly agree, up to concentrations as high as 30% by volume were carefully made avoiding aggregation. The objective was to approximate the ideal conditions under which the PSD is constant throughout the whole concentration ladder. This common PSD was determined either by an independent size measurement technique, a manufacturer specification and/or the ULTRASIZER™ at a concentration for optimal accuracy. Knowing the actual concentration at each step of the ladder and the common PSD, predictions from the METAMODEL™ were compared with the actual spectral measurements. Silica suspensions manufactured by Nissan Chemical Company with well-controlled very narrow size distributions were employed. The whole set of validation data can be found in (2).

All Models for Nissan MP1040 - Cv=30%

Practically all models available in the literature that are based on the fundamental analytical wave scattering theory have been implemented (ECAH, Foldy, Waterman & Truell (WT), Fikioris & Waterman (FW), Twersky, Lloyd & Berry (LB)). It was found, as a constant throughout the myriad of experiments conducted over the years, that all these models only display minor deviations from the single-scattering ECAH model; some towards the right direction and some towards the opposite direction with respect to the actual experimental data. These deviations are never large enough to account for the undergoing *multiple scattering* unless the concentration is below 5% (2).

Figure 1 depicts the results for a silica slurry named Nissan MP 1040 specified by the manufacturer with a size $100nm \pm 10nm$ and Cv=30%. For the sake of clarity, only the WT, FW, and METAMODEL™ are shown together with the experimental data. The rest of the models all fall within WT and FW. The actual size was determined with the ULTRASIZER™ at a concentration for maximum accuracy obtaining 105 nm. The root-mean-squared (RMS) error for Waterman & Truell model is over 360% and all the other multiple-scattering models have RMS errors over 500%. METAMODEL™ displays instead an RMS error better than 8% with typical deviations better than 5%.

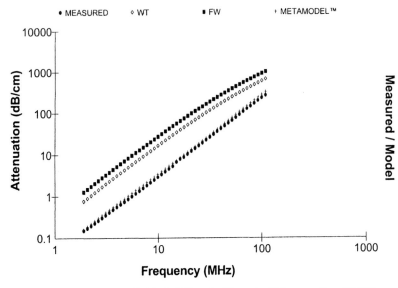

Figure 1. Nissan MP1040 (Silica in Water) -105 nm – Cv=30.0%

From this general finding, and for the sake of clarity on the plots, we will show subsequently only three spectra: the experimental, the well-known single-scattering ECAH, and the METAMODEL™. Global errors for the other models will be simply mentioned. Figure 2 shows spectra for the same Nissan MP1040 with Cv=10%. At low enough concentration, both ECAH and METAMODEL agree perfectly. However, as we can see, at a relatively low concentration of 10% they disagree drastically. ECAH global error is around 70%, while the one for METAMODEL™ is better than 5%. The best of the classical multiple-scattering models shows an RMS error no better than 50%.

METAMODEL™ Vs ECAH for Nissan MP3040

Manufacturer's specification for the MP3040 product is $300nm \pm 30nm$. The ULTRASIZER™, at the optimum concentration for best accuracy, delivered $310nm$. Figure 3 displays data for Cv=24%. The difference between experimental data and single-scattering prediction is abysmal with a global error of 190%. METAMODEL™ agreement is better than 10%.

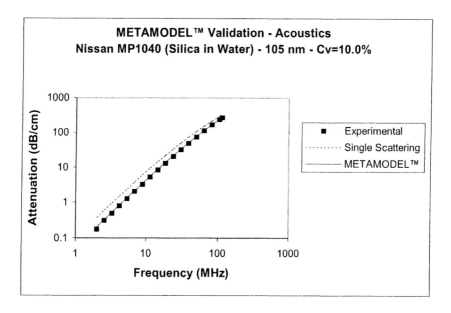

Figure 2. Nissan MP1040 (Silica in Water) - 105 nm - Cv=10.0%

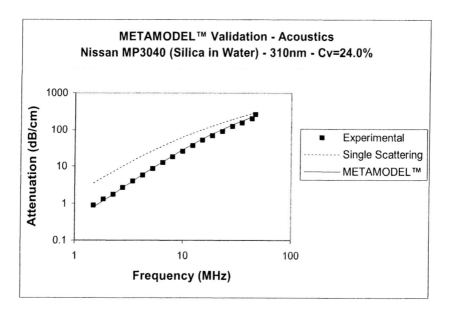

Figure 3. Nissan MP3040 (Silica in Water) - 310 nm - Cv=24%

METAMODEL™ Vs ECAH for Nissan MP4540

This Nissan product is stated to have a mean size of $450nm \pm 45nm$. At the optimal concentration, ULTRASIZER™ delivered 440 nm. Figure 4 compares theories with experimental data at Cv=10%. It is clear that at this low concentration multiple scattering is already dominant below 10 MHz. ECAH global error is around 32% while the one for METAMODEL™ is better than 5%. The best of the classical multiple-scattering models shows an RMS error no better than 23%.

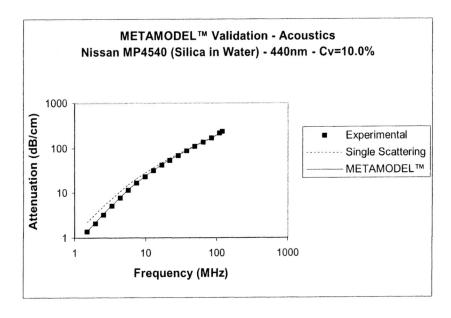

Figure 4. Nissan MP4540 (Silica in Water) -440 nm – Cv=10.0%

Figure 5 depicts the comparison for Cv=30%. Multiple-scattering is dominant with the ECAH model having an error of 260% while METAMODEL™ agreeing with the measured spectrum by better than 10% overall. The best of the classical multiple-scattering models disagrees with the experimental spectrum by more than 170%.

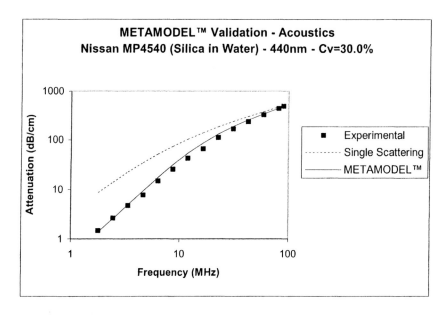

Figure 5. Nissan MP4540 (Silica in Water) - 440 nm - Cv=30.0%

Conclusions

Fundamental models for accuratetly predicting multiple scattering of electromagnetic and acoustic waves through particulates have been in need for a long time. The lack of these models has forced optical size analyzers to only accept highly diluted samples and considerably impairs the performance of acoustic instruments.

The METAMODEL™ fundamental mathematical model (2) accurately predicts (within 5%) the interaction of generic waves with concentrated particulates *opening up the possibility for light-scattering instruments to operate at concentrations several orders of magnitude higher than before.*

Similarly, METAMODEL™ *would improve dramatically the accuracy of acoustic particle size analyzers,* particularly for submicron solid particles at concentrations higher than 10% by volume.

The SCATTERER™ tool has proven extremely powerful in conducting basic research and development on the interaction of generic waves with particulates (*2*).

References

1. Allegra, J.R.; Hawley, S.A. *"Attenuation of Sound in Suspensions and Emulsions: Theory and Experiments"* J. Acoust. Soc. Am. 1972, 51, 1545-1564.
2. W*ebsite for SCATTERER™ and METAMODEL™ technologies,* URL http://www.scatterer.com.
3. Ishimaru, A. *"Wave Propagation and Scattering in Random Media"*, Academic Press, 1978.
4. Alba, F. et al *"Ultrasound Spectroscopy: A Sound Approach to Sizing of Concentrated Particulates"*, Handbook on Ultrasonic and Dielectric Characterization Techniques for Suspended Particulates, The American Ceramic Society, 1998, 111-127.
5. Varadan, V.V. et al, *"Multiple Scattering Theory for Acoustic, Electromagnetic, and Elastic Waves in Discrete Random Media"*, Multiple Scattering of Waves in Random Media and Random Rough Surfaces. The Pennsylvania State University, 1985.
6. Foldy, L.L. *"The Multiple Scattering of Waves - I. General Theory of Isotropic Scattering by Randomly Distributed Scatterers"*, Physical Review, February 1 and 15, 1945, vol 67, Numbers 3 and 4.
7. Lax, M. *"Multiple Scattering of Waves"*, Reviews of Modern Physics, October 1951,Volume 23, Number 4.
8. Waterman P.C.; Truell, R. *"Multiple Scattering of Waves"*, Journal of Mathematical Physics, July-August, 1961, Volume 2, Number 4.
9. Stratton, J.A. *"Electromagnetic Theory"*, McGraw-Hill Book Company, Inc. 1941.

Chapter 11

Toward a Model for the Mobility of Circularly Permuted DNA Fragments in Gel

Aleksander Spasic[1], Kevin Yang[2], and Udayan Mohanty[1]

[1]Department of Chemistry, Boston College, Chestnut Hill, MA 02467
[2]Research Science Institute, Massachusetts Institute of Technology, Cambridge, MA 02139

A model is proposed that is capable of quantitatively describing the gel electrophoretic mobility patterns of circularly permuted B-DNA fragment. The model takes into account in an approximate way the interaction of the DNA with the gel matrix thorough an effective elastic force constant. The results are compared with experimental data and with the reptation model.

150

Introduction

The migration patterns of DNA fragments in polyacrylamide gel is governed exquisitely by the shape of the molecule (1). This suggests that comparative electrophoresis, i.e., electrophoresis measurement of DNA constructs relative to straight DNA fragments of identical length can be suitably exploited to unravel the relative magnitude, the position, and perhaps even the direction of the bend along the helical backbone of macromolecule.

Comparative gel electrophoresis was first exploited to identify and characterize some of the sequence elements responsible for bending of DNA molecules (2-6). The basic idea is illustrated in Figure 1. Here, two different restriction enzymes cleave a tandem dimer. The two cleaved DNA fragments are not only circularly permuted versions of the same sequence but are also of equal length. However, one of the fragments has a bend near its end, while the other one fragment has a bend at the center. Since a fragment with a bend at the center has a more overall curvature than the molecule with bends near the end, the former is expected to migrate slower due to entanglement with the gel fibers (1).

What is the molecular origin for the reduction in mobility due to intrinsic curvature in DNA? According to conventional wisdom, large DNA fragments reptate through polyacrylamide gels (7-12). The gel fibers impose restraining forces on the DNA. These forces lead to electrophoretic migration of the polyion through tubelike regions. The electrophoretic mobility μ of the DNA is related to its effective charge Q, the translation frictional coefficient ζ along the tube,

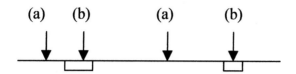

Tandem dimer

(a) construct has low mobility   (b) construct has high mobility

Figure 1. A tandem dimer is cleaved by two different restriction enzymes and produces constructs (a) and (b). The rectangular boxes denote the position of the bends. Construct (a) will migrate slowly in polyacrylamide gel relative to construct (b) due to more overall curvature.

the contour length L, and the end-to-end vector h_x of the polyion in the tube projected along x-direction -- the direction of the external electric field (8,10)

$$\mu = (Q / \zeta) < h_x^2 / L^2 >. \tag{1}$$

The angular brackets denote an average over DNA conformations. If the shape of the DNA in the gel mirrors the shape observed in free solution, then $< h_x^2 / L^2 >$ is proportional to the mean squared end-to-end distance of the molecule. Eq. (1) predicts that a DNA construct with a bend at the center has smaller mean squared end-to-end distance and hence would migrate slower than a molecule of identical length but with a bend near the end.

In this paper we propose a model that quantitatively describes the gel electrophoretic mobility of oligomeric B-DNA fragment with a single bend at the center. The essential features of the model are described in the next section. The predictions of the model are compared with experimental data and with the reptation model in the following section.

Model

In this section we describe the essential ingredients of a model that enable us to quantitatively describe the electrophoretic mobility of B-DNA with a variable bend at the center. The details of the derivation will not be discussed here (13).

In the presence of an external electric field E, the DNA in aqueous solution experiences a force that is the product of its effective charge Q and the electric field. The frictional force that retards the motion of the DNA is the product of the frictional coefficient λ and its velocity V. If steady state condition prevails, as it does in free solution electrophoresis experiments, the two forces balance. Since the free solution mobility μ_O of DNA is experimentally deduced from the ratio of its velocity to the external electric field, the mobility is expressible in terms of the ratio of the effective charge and the friction coefficient (14).

To determine the free solution mobility of DNA requires a discussion of the various factors that govern the frictional coefficient λ and the effective charge of the molecule. Let us first focus on the friction coefficient.

A charge in motion produces an electric field. The total electric field at a spatial point in the solution and at a given time results from the ionic charge densities and the electric field of the moving charges; the later satisfies the Poisson equation. The ionic densities are governed by diffusion like equation. The ionic flux has contributions from diffusion, convection and conduction and is coupled to the hydrodynamic equation via the electric potential and the velocity of the liquid (15,16). The characteristics of distortion of the Debye cloud, often coined the asymmetrical field effect, due to moving charge is obtained by solving the Poisson equation with the time variation of the ionic densities in the steady-state approximation. On linearizing with respect to the velocity of the polyion, an expression for the ionic charge densities is obtained from which the resulting electric field acting on a monomer in aqueous solution is deduced (15,16).

A further factor contributing to the friction is the solvent that flows past the moving polyion during electrophoresis. These flows induce long ranged hydrodynamic interactions between distant monomers along the backbone of the polymer chain. To determine this effect we consider the various forces that act on a monomer such as that due to Brownian motion, the polarization forces due to distortion of the Debye cloud as well as a body force term (15,16). The later term is proportional to the product of the charge density and the electric field.

Now consider the net effect of the inclusion of the body force term in the hydrodynamic equation. If one assumes steady state conditions, then the divergence of the ionic flux vanishes. The spatial distribution of ions and the local liquid velocity are constant and this leads to conservation of ion transport. The concentration of cations and anions is expanded around the distributions of ions in the absence of the electric field. The unperturbed distribution of the small ions is expressed in terms of the product of bulk salt concentration and the Boltzmann factor of the ratio of the product of the electronic charge and the potential of the unperturbed double layer to the thermal energy at temperature T. The unperturbed double layer potential satisfies the Poisson-Boltzmann equation.

Analysis reveals that the hydrodynamic velocity field is proportional to the external field times the effective charge of the macromolecule, the proportionality being the screened Oseen tensor (15,16,17). The decay of the Oseen tensor with distance is governed by the Debye screening parameter κ, which for monovalent ions is given by $\kappa = I^{1/2}/3$ in units of inverse Angstrom and I is the ionic strength in molarity.

Additional factor contributing to DNA mobility is the restraining force that it experiences due to the gel fibers. The restraining forces change the velocity of the DNA relative to that in free solution and leads to two consequences. First, the effective charge is renormalized relative to that in free solution (18). Second, the velocity of the DNA is reduced proportional to the exponential of the size R of the DNA relative to the average mesh size χ (19-21). This can be justified based on linearized hydrodynamic equation for the diffusion of spherical molecules in the presence of a random distribution of fixed, rod like molecules of a given density (22). Experimental studies have also established that the product of their pure solution diffusion coefficient and an exponential function of the ratio between the size of the molecule and the average mesh size give the diffusion of small macromolecules in polyacrylamide gels (23).

We now turn our attention to the determination of the effective charge of the DNA. It is well established that under appropriate conditions counterions from aqueous solution condense on the DNA and neutralize a fixed fraction of the bare charges on the backbone (24,25). This is the so-called Manning counterion condensation phenomenon (24,25). Detailed Monte Carlo and molecular dynamics simulations by Jayaram *et al.* and Mills *et al.* on various structural forms of DNA indicate that the counterions form a delocalized cloud that shields the negative phosphate charges from the ions in the solvent (26). After counterion condensation, there still remain the unneutralized charges. The counter- and the co-ion clouds screen the charges from each other (27). The total fraction of counterions per polyion charge is a sum of the condensed and the screened counterions and can be expressed solely in terms of the liner charge density parameter defined as the ratio of the Bjerrum length l_B and the distance b between the charges (28).

The various assumptions lead to an expression for the electrophoretic mobility

$$\mu = \mu_o \frac{\overline{R}_{end}}{L} e^{-F(\theta)/k_BT} e^{-R/\chi},\qquad(2)$$

$$\mu_O = \frac{Q/6\pi\eta}{\displaystyle\sum_{i}^{N}\sum_{j\neq i}^{N} <exp(-\kappa r_{ij})/r_{ij}>}\frac{1}{1-T_o}.\qquad(3)$$

Here, N is the number of monomers, the angular brackets represent an average over the conformation of the polymer chain, r_{ij} is the distance between monomer i and monomer j and η is the solvent viscosity. $F(\theta)$ is the free energy of bending a polyion through an angle θ. The term T_O is proportional to $<\sum_{ij} e^{(-\kappa r_{ij})}>$ and comes from accounting for asymmetrical field effects

(15,16). $\overline{R}_{end}$ is the square root of the mean squared end-to-end distance and can be expressed in terms of the persistence length P (29)

$$\overline{R}_{end} = \sqrt{\begin{array}{l} 2PL\{1 - (P/L)[(1 - e^{-L/2P}) + (1 - e^{-L/2P}) \\ - \cos\theta(1 - e^{-L/2P})(1 - e^{-L/2P})]\} \end{array}} \, .$$

(4)

For a straight DNA fragment Eqs. (2)-(3) reduces to that proposed by Mohanty and coworkers (14,19-21).

Since the experimentally relevant quantity is the ratio of the mobility of a bend DNA to that of a straight DNA of identical length, what enters our model is the difference in free energy between a bend and a straight DNA, $\Delta F(\theta) = F(\theta) - F(0)$, where ΔF is given by

$$\Delta F = B_{eff}(\Delta\theta)^2 .$$

(5)

B_{eff} is an effective bending force constant for the combined DNA-gel system (10).

Results and Discussion

In this section we describe the various structural parameters of the macromolecule and the characteristic of the buffer that enter the model. The predictions for electrophoretic mobility of the curved B-DNA are then compared with available experimental data and the reptation model.

Buffer and structural characteristics of ionic oligomer

The quantity R in Eq. (1) is the effective size or radius of the migrating macromolecule. There are three standard ways to estimate R. For polymer chains with cylindrical symmetry, the molecular size can be estimated from the root mean-squared radius of gyration (30). Another estimate is to use the geometrical mean value radius; it is defined as the radius of the sphere whose volume is equal to the volume of the cylindrical ionic oligomer (30). This last approach, the path that we followed, is to estimate the size of the macromolecule by the relation $R = bN / 2$. The qualitative features of our predictions for the electrophoretic mobility do not change on making use of the remaining two alternative ways.

The structural features of B-DNA entering the model are the number N of charges and the spacing b between the charges. The charge spacing $b = 1.7$ Å and the Bjerrum length is 7.1 Å. The contour length of the DNA is 301 bp corresponding to the Thompson-Landy construct (4). A bend of magnitude θ is introduced at the center of the DNA as shown in Figure 2. The bend angle is varied from zero to 130 degrees.

The "bare" elastic modulus B_0 of the DNA is a function of the persistence length and the contour length. The persistence length is taken to be

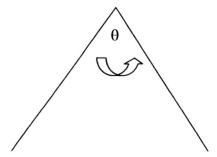

Figure 2. A schematic representation of a DNA bent at the center. The angle of bend is denoted by θ.

530 Å. Assume that the effective elastic modulus is proportional to the "bare" elastic modulus, $B_{eff} = \alpha B_O$, where α is a parameter that is governed by the properties of the gel and the nature of its interaction with the bent DNA. We estimate this parameter as follows. In a dynamic Monte Carlo simulation, Levene and Zimm elucidated the anomalous migration of bent DNA molecules by taking into account the elastic free energy of the DNA (10). To account for the discrepancy between the simulated mobility and experimental data, it was assumed that the effective elastic modulus is approximately a factor of ten smaller than that determined from values based on persistence length and the average mesh size (10).

This value is consistent with that obtained by taking into account the interactions of the gel fibers with the DNA monomers in an approximate way; in this case, the parameter α can be related to surface charge density of the DNA, the Bjerrum length, the charge spacing, and the Debye screening parameter (28). Thus, we have varied α between 0.1 and 0.2.

The buffer typically used in free solution and gel experiments is TBE. It consists of 45 mM Tris borate and 1 mM EDTA with pH equal to 8.4. The ionic strength of the buffer deduced from the Henderson-Hasselbach equation is 0.0256 M (15).

Experimental mobility measurements of DNA are carried out in 8% polyacrylamide gel with *29 : 1* acrylamide-bisacrylamide cross-link ratio. The corresponding mesh size χ is approximately in the range *70 – 100* Å (31).

Comparison of the model to experimental data

A comparison of the relative mobility of bend DNA with respect to straight DNA of identical length as a function of angle of bend is shown in Figure 3. The predictions are in good agreement with the experimental data of Thompson and Landy (4).

To test the predictions of the reptation model with the experimental data is not possible in view of the fact that the translation frictional coefficient ζ along the tube axis is not a measurable quantity. However, if we make the hypothesis that the ratio Q / ζ appearing in Eq. (1) can be identified with the free

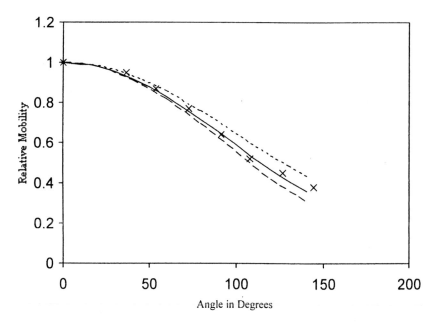

Figure 3. The mobility of a bend DNA with respect to a straight DNA as a function of the bend angle is compared with the Thompson-Landy data (4) and with the reptation model. The crosses represent the experimental data, while the doted (α =0.1) and the continuos (α =0.16) curves are the theoretical prediction of our model for two values of the effective bending constant as reflected by the quantity α. The dashed line is the reptation model appropriately modified as discussed in the text.

solution mobility of the DNA, then a comparison with experimental data is indeed feasible. This is so because if the conformation of the DNA in the gel reflects that which is observed in aqueous solution, then $< h_x^2 / L^2 > \approx < h_x^2 > / L^2$. But the mean squared end-to-end distance can be numerically calculated given the charge spacing of the phosphate groups and the geometry of the DNA construct (Figure 2). The prediction of the reptation model for the relative mobility is also depicted in Figure 3.

Several comments are in order. First, the mean spacing of the fibers in the gel is much less than the persistence length implying an apparent "tight" gel matrix. Second, since α is less than unity, it suggests that the elastic free energy of the DNA is a result of its interaction with an elastic gel matrix. Thus, an improved model for mobility must account for the viscoelastic response of the gel (32). Third, bending generally occurs over several nucleotides (1-6). Hence, modeling intrinsic curvature of DNA with a bend at a single position along its backbone has shortcomings.

Concluding remarks

In this paper we have presented a model capable of quantitatively describing the gel electrophoretic mobility patterns of circularly permuted B-DNA fragment. Our model takes into account in an approximate way the interaction of the DNA with the gel matrix thorough an effective elastic force constant. The results are in good agreement with experimental data.

References

1. Crothers, D. M.; Drak. *Methods in Enzymology*, **1992**, 212, 46.

2. Wu, H. M. ; Crothers, D. M. *Nature (London)* **1984**, 308, 509.

3. Koo, H. S.; Wu, H. M.; Crothers, D. M. *Nature (London)* **1986**, 320, 521.

4. Thompson, J. F.; Landy, A. *Nucl. Acids Res.* **1988**, 16, 9687.

5. Bolshoy, A.; McNamara, P.; Harrington, R. E.; Trifonov, E. N. *Proc. Nat. Acad. U.S.A.* **1991**, 88, 2312.

6. Ulanovsky, L.; Trifonov, E. N. *Nature (London)* **1987**, 326, 720.

7. de Genne, P. G. *J. Chem. Phys.* **1972**, 55, 572.

8. Zimm, B. H.; Lumpkin, O. J. *Macromolecules* **1993**, 26, 226.

9. Lumpkin, O. J.; Levene S. D.; Zimm, B. H. *Phys. Rev. A.* **1989**, 39, 1573.

10. Levine, S. D.; Zimm, B. H. *Science.* **1991**, 94, 396.

11. Lumpkin, O. J.; Dejardin, P.; Zimm, B. H. *Biopolymers* **1985**, 24, 1573.

12. (a) Perkins, T. T.; Smith, D. E.; Chu, S. *Science* **1994**, 264, 819. (b) Smith, D. E.; Perkins, T, T.; Chu, S. *Phys. Rev. Lett.* **1995**, 75, 4146.

13. Mohanty, U. unpublished, **2002**.

14. Mohanty, U.; McLaughlin, L. W. *Ann. Rev. Phys. Chem.* **2001**, 52, 93.

15. Mohanty, U.; Stellwagen, N. C. *Biopolymers* **1998**, 49, 209.

16. Barrat, J. L. and Joanny, J. F. **(1996)**. In Prigogine, I., and Rice, S. A. (eds.), Adv. Chem. Phys. John

    Wiley &Sons, Inc., Vol. XCIV, pp. 1-66.

17. Manning, G. S. *J. Phys. Chem.* **1981**, 85, 1506.

18. Stigter, D. *Biopolymers* **1991**, 31, 169.

19. Mohanty, U.; Searls, T.; McLaughlin, L. W. *J. Am. Chem. Soc.* **1998**, 120, 8275.

20. Mohanty, U.; Searls, T.; McLaughlin, L. W. *J. Am. Chem. Soc.* **2000**, 122, 1225.

21. Mohanty, U.; Searls, T.; McLaughlin, L. W. *J. Biomol. Struct.* Dyn. **2000**,1, 371.

22. Cukier, R. I. *Maxcromolecules* **1984**, 17, 252.

23. Cobbs, G. *Biophys. J.* **1981**, 35, 535; Park. Il. H.; Johnson, C. S.; Gabriel, d.

    A. Macromolecules **1990**, 23, 1548.

24. Manning, G. S. *Q. Rev. Biophys.* **1978**, 11, 179.

25. Fenley, M. O.; Manning, G. S.; Olson, W. *Biopolymers* **1990**, 30, 1191.

26. (a) Jayaram, B.; Swaminathan, S.; Beveridge, D. L. *Macromolecules* **1990**, 23, 3156. (b) Mills, P. A.; Rashid, A.; James, T. L. *Biopolyemrs* **1992**, 32, 1481.

27. Record, M. T.; Lohman, T. M.; de Haseth, P. de. *J. Mol. Biol.* **1976**, 197, 145.

28. Barrat, J. L.; Joanny, J. F. J. Phys. II France **1992**, 2, 1531.

29. Rivetti, C.; Walker, C.; Bustamante, C. J. Mol. Biol. **1998**, 280, 41.

30. Stellwagen, N.C. *Biochemistry* **1983**, 22, 6186.

31. Stellwagen, N. C. *Electrophoresis* **1997**, 18, 33.

32. Mohanty, U.; Taubes, C. H. unpublished, **2002**.

Chapter 12

Viscoelastic Properties of Solid–Liquid Dispersions

Tharwat Tadros

89 Nash Grove Lane, Wokingham, Berkshire RG40 4HE, United Kingdom

This overview desribes the viscoelastic properties of concentrated solid/liquid dispersions (suspensions). After a brief description of the basic principles of viscoelastic measurements, four systems were described. Hard-sphere dispersions, electrostatically stabilised dispersions, sterically stabilised dispersions and flocculated (weak and strong) dispersions. An attempt was made to correlate the viscoelastic properties with the interparticle interactions.

The present review is aimed at describing the rheological properties of suspensions and attempting to relate these properties to the interparticle interactions. As we will see later suspensions may show viscous, elastic or viscoelastic (mixed viscous and elastic) response depending on the ratio between the relaxation time of the system (which depends on its volume fraction, particle size, hydrodynamic and interparticle interaction) and the experimental time. A brief description of viscoelasticity will be first given and this is followed by various sections on the rheology of four main classes of suspensions : Hard-sphere, electrostatic, steric and flocculated (both weak and strong) systems.

PRINCIPLES OF VISCOELASTIC MEASUREMENTS

For full evaluation of the viscoelastic properties of concentrated dispersions one needs to carry out experiments within well defined time scales. Two main types of experiments may be carried out. In the first type, referred to as transient measurements, a constant stress or strain is applied within a well defined short period and the relaxation of strain or stress respectively is followed as a function of time. In the second type, referred to as dynamic measurements, a stress or strain is applied within a well characterized frequency regime (usually sinusoidal) and then resulting strain (or stress) is compared with the stress (or strain) respectively. Although the two types of measurements are equivalent, they provide different information.

The above mentioned experiments are usually referred to as low deformation since measurements can be made before the "structure" of the suspension is broken down. These low deformation experiments provide valuable information of the "structure" of the suspension at any interparticle interaction. However, experiments can also be carried under conditions whereby the "structure" of the suspension is deformed or broken down during the measurement. This is the case, for example, whereby the suspension is subjected to continuous shear while measuring the stress in the sample. Such measurements are sometimes referred to as steady state (high deformation) measurements. Below a summary of the basic principles of the above mentioned measurements is given.

Transient (Static) Measurements

As mentioned above one can either measure the relaxation of stress after sudden application of strain or the strain relaxation after sudden application of stress (creep measurements). In stress relaxation a sudden shear strain γ is applied on the system within a very short shear period, i.e. keeping the rate of strain γ constant (1). The stress decays exponentially with time for a system with a single relaxation time that follows a simple Maxwell model. The ratio of the stress at any time t to the constant strain applied is the stress relaxation modulus G(t). As the stress in the sample relaxes by viscous flow, the modulus decreases. The instantaneous value of the stress at the moment when the strain is applied is τ_0 and the corresponding modulus is G_0. The stress relaxes exponentially with time and reaches a zero value at $t = \infty$. The system is referred to as a viscoelastic liquid. In this case, the rate of decrease of stress with time follows an equation similar to first order kinetics (1). The stress relaxation modulus is given by the equation,

$$G(t) = G_0 \exp(-t/t_r) \qquad (1)$$

where t_r is the relaxation time that is given by the ratio of the viscosity to the modulus.

As mentioned above, for a viscoelastic liquid the modulus becomes zero at infinite time. However, for a viscoelastic solid, the modulus reaches a finite equilibrium value at $t = \infty$ and in this case one of the relaxation times is infinite and the equilibrium modulus is G_e.

The second type of transient measurements is the constant stress or creep. In this case, a small stress τ is applied on the system and the strain or compliance J (deformation per applied stress, i.e. γ/τ) is followed as a function of time. At time t, the stress is suddenly removed and the deformation, which now reverses sign, is measured over a longer period of time. At $t = 0$, there will be a rapid elastic deformation (i.e. no energy is dissipated within such a very short time period), characterized by an instantaneous compliance J_o that is inversely proportional to the instantaneous modulus G_o ($G_o = \tau/J_o$). At $t > 0$, J increases less rapidly with time and the system exhibits a retarded elastic response.

The form of the creep curve depends on whether the material behaves as a viscoelastic liquid or solid. This is particularly important after sudden removal of stress. The rate of deformation will change sign and the system will return more or less towards its normal state. This reversing of deformation is called creep recovery. With a viscoelastic solid , the system reaches an equilibrium deformation in creep, characterised by an equilibrium compliance J_e. For a viscoelastic liquid, on the other hand, such as a dispersion where the particles are not permanently attached to each other, the system does not reach an equilibrium compliance. Under constant stress, the strain rate approaches a limiting value and a situation of steady flow is eventually reached, governed by a Newtonian viscosity η_o.

Dynamic (Oscillatory) Measurements

In these experiments, a small amplitude sinusoidal strain (or stress) with frequency v (Hz) or ω (rad s^{-1}), where $\omega = 2\pi v$, is applied to the system and the stress and strain compared simultaneously (1). The amplitude of the stress is τ_o and it oscillates with the same frequency, but out of phase. The phase angle shift δ is given by the product of the time shift Δt between the strain and stress sine waves and the frequency ω. For a perfectly elastic system, the maximum stress occurs when the strain is a maximum. In this case $\delta = 0$. For a perfectly viscous liquid, the maximum stress occurs at the maximum strain rate, and the two waves are 90° out of phase.

Dynamic measurements are usually expressed in terms of a complex modulus, $|G^*|$ that is given by,

$$|G^*| = \frac{\tau_o}{\gamma_o} \qquad (2)$$

The complex modulus is vectorially resolved into two components, the elastic or storage modulus G' (the real part of the complex modulus) and the viscous or loss modulus G'' (the imaginary part of the complex modulus), i.e.,

$$G' = |G^*| \cos\delta \tag{3}$$

$$G'' = |G^*| \sin\delta \tag{4}$$

G' is a measure of the energy stored elastically during a cycle, whereas G'' is a measure of the energy dissipated as viscous flow. It is convenient to define the ratio of G'' to G',

$$\tan\delta = \frac{G''}{G'} \tag{5}$$

In dynamic measurements, one initially measures the variation of G^*, G' and G'' with strain amplitude, at a fixed frequency, to obtain the linear viscoelastic region where the rheological parameters are independent of the applied strain amplitude. Once this linear region is established, measurements are then made at a fixed strain amplitude (within the linear region) as a function of frequency.

VISCOELASTIC PROPERTIES OF CONCENTRATED SUSPENSIONS

The viscoelastic properties of suspensions is determined by the balance of three main forces: Brownian diffusion, hydrodynamic interaction and interparticle forces. The latter are the double layer repulsion, the van der Waals attraction and steric interaction. The range of interaction is determined by the volume fraction, ϕ, and the particle size (radius R) and shape distribution. The rheology of concentrated suspensions is complex and various responses may be obtained depending on the time scale of the experiment and the structure of the system (which determines its relaxation time). In this respect, it is useful to use the ratio of the relaxation time, t_r, to the experimental time, t_e, as a means of classification of the various responses. This dimensionless group is referred to as the Deborah number, D_e. i.e.,

$$D_e \; = \; \frac{t_r}{t_e} \qquad (6)$$

If $D_e >> 1$, one obtains elastic deformation and the response is described as "solid-like" behaviour. This, for example, is the case when measurements are made at high frequency (t_e is small) for a system with high relaxation time. When $D_e < < 1$, one obtains viscous deformation and this is referred to as "fluid-like" behaviour. This occurs at low frequencies (t_e is large) for systems with low relaxation times. Many colloidal systems show a viscoelastic response with a D_e in the region of 1, whereby the experimental time scale of the measurement is comparable to the relaxation time of the system. Thus, by varying the time scale of the experiment (such as frequency in dynamic measurements) one can obtain various responses and the results may be correlated with the various interaction forces in the system. Alternatively, one can use a fixed time scale regime (a fixed frequency range) and vary some parameters of the system such as volume fraction, particle size and surface forces to obtain information on the interaction in the system. This will be illustrated below.

Four different systems may be distinguished: hard-sphere suspensions, electrostatically stabilized suspensions (soft interactions), sterically stabilized dispersions and flocculated (and coagulated) systems. These systems increase in the order of the complexity of their rheology, with the hard-sphere systems being the most simple and the flocculated or coagulated systems being the most complicated both experimentally and theoretically. Progress on the rheology of concentrated suspensions has been slow and only in recent years some advances have been made. This is due to the development of modern rheological techniques. However, theoretical analysis of the rheological data is far from being quantitative and in most cases the theories are based on many approximations. Inspite of these limitations, viscoelastic measurements provide valuable information on the interactions in concentrated suspensions and in some cases it is possible to obtain the magnitude of the forces involved.

Suspensions with Hard-Sphere Interactions

These are sometimes referred to as systems with neutral stability in which both repulsion and attraction are screened. In other words, all interactions are weak and the main forces responsible for flow are hydrodynamic interaction and Brownian diffusion. The appropriate dimensionless group is $\gamma \times t_r$, where γ is the shear rate and t_r is the time scale of Brownian diffusion.

Model dispersions of hard spheres have been developed by Krieger and his collaborators (2,3) who used monodisperse polystyrene latex dispersions with well characterised surface and particle radius. To minimise repulsion, the double layer was compressed either by addition of electrolyte or by replacement of water with a less polar medium such as benzyl alcohol. Under these conditions, both attraction and repulsion are minimized and the dispersion behaves as a hard-sphere system. To check this, it is essential to plot the relative viscosity η_r versus the reduced shear rate (or reduced shear stress) at a particular volume fraction, with several particle radii and fluid viscosities to see if the relationship obeys a rheological equation of state of the from (2,3),

$$\eta_r = f(\phi, \gamma_r) \tag{7}$$

All viscosity data for the various particle size latices fall on the same line indicating that the systems behave as hard-sphere suspensions. expected. If η_r is plotted versus ϕ, a general behaviour is obtained as illustrated in Fig.1 whereby the high shear relative viscosity for polystyrene latex suspensions is plotted versus ϕ. The relative viscosity increases gradually with increase of ϕ, but above a certain ϕ value it shows a rapid increase with further increase of ϕ reaching an asymptote when ϕ reaches about 0.6. The slope of the curve in the limit $\phi \rightarrow 0$, gives the intrinsic viscosity, which for hard-spheres is equal to 2.5, whereas the value of ϕ at the asymptote represents the maximum packing fraction, ϕ_p, for the hard-sphere suspension (which is of the order of 0.64 for random packing). The relative viscosity - volume fraction curve shown in Fig.1 could be represented by the following equation (2,3),

$$\eta_r = \left[1 - \left(\frac{\phi}{\phi_p} \right) \right]^{-[\eta]\phi_p} \tag{8}$$

Equation (8) is usually referred to as the Dougherty-Krieger equation.

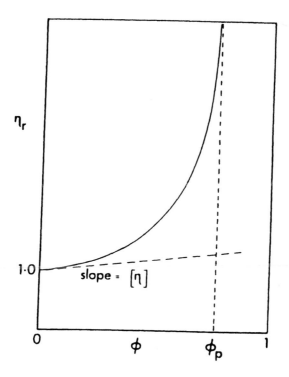

Fig.1 Typical plot of η_r versus ϕ

A theory for the rheology of hard-sphere dispersions has been developed by Bachelor (4) who considered the balance between hydrodynamic interaction and Brownian diffusion, i.e.,

$$\eta_r = 1 + 2.5 \phi + 6.2 \phi^2 + 0 \phi^3 \tag{9}$$

The first two terms on the right hand side of equation (9) represent the

Einstein's limit, whereas the third term (6.2 ϕ^2) is the contribution from hydrodynamic interaction; the third term in ϕ^3 represents higher order interactions. Bachelor's theory is valid for $\phi < 0.2$. Extension of the above theory to more concentrated suspensions requires the introduction of higher order interactions. No theory is, as yet, available that treats such a complex problem. Computer simulation of the multibody interaction may offer a starting point for predicting the viscosity of concentrated suspensions.

Stable Systems with Soft (Electrostatic) Interactions

These are systems with expanded double layers, i.e. at low electrolyte concentrations, whereby the interaction is dominated by double layer repulsion. The appropriate dimensionless group characterizing the process (balance of viscous and repulsive force) is given by $\eta_o a^2 \gamma / \varepsilon \psi_o^2$, where ψ_o is the surface (or zeta) potential. The viscoelastic properties of these systems can be investigated using constant stress (creep) or oscillatory measurements. As an illustration, Fig.2 shows plots of G^*, G' and G'' (at one frequency, namely 1 Hz and low strain) versus the volume fraction ϕ (the core volume fraction, i.e. excluding the contribution from the double layer) for a latex suspension with a radius of 700 nm (5). A logarithmic scale is used for the moduli values. All the results were obtained at low strain values to be as close as possible to the linear viscoelastic region. In 10^{-5} mol dm^{-3} NaCl, the moduli values show a rapid increase with increase in ϕ within the range studies, namely 0.460 - 0.524. In addition, G' is always greater than G'' within this volume fraction range. At the higher ϕ value (0.524) G' approaches G^* very closely and the suspension behaves as a near elastic solid. In contrast, the results in 10^{-3} mol dm^{-3} NaCl, show a rapid increase in G' and G'' when ϕ exceeds 0.53. In addition, within the volume fraction range studied (0.3 - 0.566) G' is either close to G'' or even lower than it.

The above trends can be adequately explained if one considers the presence of the double layer around the particles. To a first approximation, the double layer thickness, $1/\kappa$, should be added to the particle radius to obtain the effective radius, a_{eff}. At any given size, a_{eff} depends on the electrolyte concentration, C, since the double layer thickness is determined by C. For example, in the above case a_{eff} is ~ 800 nm in 10^{-5} mol dm^{-3} NaCl, whereas a_{eff} is ~ 710 in 10^{-3} mol dm^{-3} NaCl. Since the volume fraction scales with the cube of the radius, it is clear that, at any given ϕ, the effective volume fraction of the suspension is much higher at the lower electrolyte concentration. ϕ_{eff} is related to ϕ by the following expression,

$$\phi_{\text{eff}} \;=\; \phi \left[1 + \frac{(1/\kappa)}{a} \right]^{3} \qquad\qquad \textbf{(10)}$$

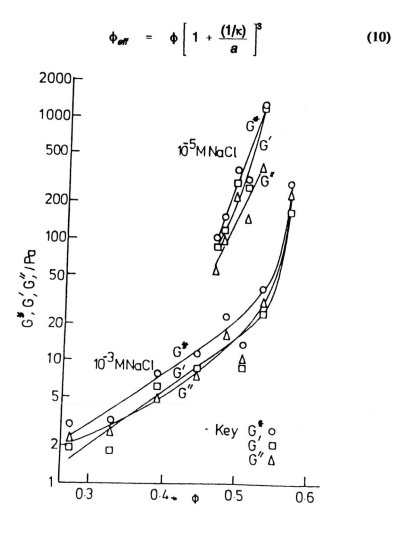

Fig.2 Variation of G^{*}, G' and G'' (at $\nu = 1$ Hz) with ϕ at two electrolyte concentrations; o, G^{*}; □, G'; △, G''.

In 10^{-5} mol dm^{-3} NaCl, $\phi_{\text{eff}} = 1.5\ \phi$. Thus, at the lowest ϕ value studied, namely 0.46, $\phi_{\text{eff}} = 0.7$, which is above the maximum packing fraction (0.64 for random packing). Under these conditions, the double layer interaction is significant and some overlap of the double layers may occur. This explains why the response at this electrolyte concentration and volume fraction is predominantly elastic. At the highest ϕ value studied, namely 0.524, $\phi_{eff} = 0.79$, which is significantly higher than the maximum packing fraction. This results in significant double layer overlap and the dispersion behaves as a near elastic solid with $G' \sim G^*$. In 10^{-3} mol dm^{-3} NaCl, however, $\phi_{eff} = 1.05 \phi$ and up to the maximum volume fraction studied (0.565), ϕ_{eff} is well below the maximum packing fraction. This implies that there is insignificant overlap of the double layers and the dispersions show more viscous than elastic response. To achieve significant elastic response one has to increase the volume fraction to values above 0.6.

Sterically Stabilized Suspensions

These are suspensions where particle repulsion results from interaction between adsorbed or grafted layers of nonionic surfactants or polymers. The appropriate dimensionless group characterizing the flow (balance of viscous and steric repulsive forces) is given by $\eta_o R^2 \gamma V_s / (1/2 - \chi) \delta^2$, where χ is the Flory-Hugguns interaction parameter and δ is the thickness of the adsorbed layer. Steric interaction is repulsive as long as $\chi < 0.5$. With short chains, the interaction may be represented by a hard-sphere type with an effective radius $a_{eff} = a + \delta$. This is particularly the case with non-aqueous suspensions with an adsorbed layer that is small compared to the particle radius,. and where any electrostatic interaction is negligible. The rheology of such suspensions approach the hard-sphere behaviour. Results on aqueous sterically stabilized suspensions were obtained by Tadros and collaborators (6,7). Polystyrene latex suspensions with grafted polyethylene oxide (PEO) chains (M = 2000) were prepared using the Aquersymer process (8). As an illustration, Fig.3 shows the variation of G^*, G' and G'' with frequency ω in Hz for latex suspensions at various volume fractions. At $\phi = 0.44$, $G'' >> G'$ and all the moduli values are very low. This reflects the relatively weak interaction between the particles at such low volume fraction, since the surface-to-surface separation between the particles is larger than twice the grafted polymer layer thickness. When ϕ is increased to 0.465, G'' is still higher than G', but the moduli values increase by about a factor of 2 compared to the values at $\phi = 0.44$. As the volume fraction is increased, the steric interaction between the particles increases, since the surface-to-surface distance between the particles becomes smaller. The surface-to-surface distance between the particles at $\phi = 0.465$ is still larger than 2δ and hence the dispersion shows a predominantly viscous response within the frequency range studied. When ϕ is increased to 0.5, G' becomes now larger than G'' within the frequency range studied and the moduli values are increased by an order of magnitude compared to the values obtained at $\phi = 0.44$.

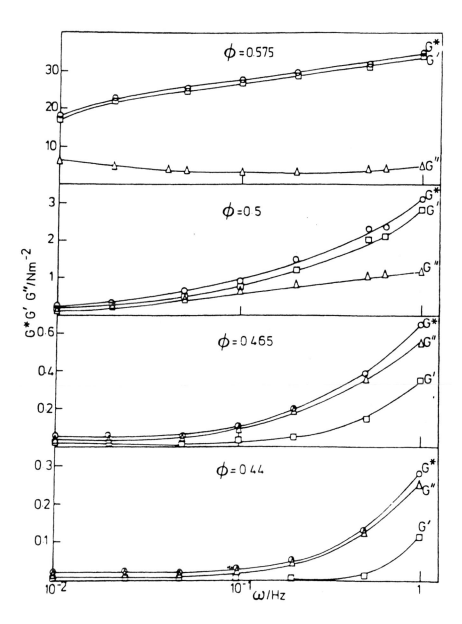

Fig.3 Variation of G*, G' and G'' with ω (Hz) for polystyrene latex suspensions (a = 175 nm), containing grafted PEO chains, at various volume fractions.

At $\phi = 0.5$, the interparticle distance (assuming random packing) is of the order of 30 nm, which is now smaller than 2δ and elastic interaction between the grafted chains becomes strong. As the volume fraction is further increased above 0.5, such elastic interaction becomes stronger and stronger and ultimately the suspension behaves as a near elastic solid. This is illustrated by the results at $\phi = 0.575$, which shows that $G' > > G''$ (and G' becomes close to G^*) and the moduli show much less dependence on frequency. The interparticle separation distance at such volume fraction is of the order of 12 nm, which is significantly smaller than 2δ. This results in interpenetration and/or compression of the grafted PEO chains resulting in very strong elastic interaction.

The correlation of the rheology of concentrated sterically stabilized dispersions with interparticle interactions has been recently investigated by Costello et al (9-12). Basically, one measures the energy E(D)-distance D curves for a graft copolymer consisting of poly(methyl methacrylate) backbone with PEO side chains (with similar molecular weight as that used in the latex) which is physically adsorbed on smooth mica sheets. The forces between mica surfaces bearing the copolymer are converted to interaction potential energy using the Deryaguin approximation for cross cylinders (13). Using de Gennes scaling theory (14), it is possible to calculate the energy of interaction between the polymer layers. The high frequency modulus can be calculated from the potential of mean force and the thoretical values are compared with the experimental results. The trends in the variation of the high frequency modulus with the volume fraction of the latex dispersions agreed well with the theoretical calculates based on the energy-distance measurements (9 - 12).

Flocculated and Coagulated systems

The rheology of unstable systems poses problems both from the experimental and theoretical points of view. This is due to the non-equilibrium nature of the structure at rest, resulting from the weak Brownian motion (15). For this reason, advances on the rheology of suspensions, where the net force is attractive, have been slow and only of qualitative nature. On the practical side, control of the rheology of flocculated and coagulated systems is difficult, since the rheology depends not only on the magnitude of the attractive forces but also on how one arrives at the flocculated or coagulated structure in question. Various structures may be formed, e.g. compact flocs, weak and metastable structures, chain aggregates, etc. At high volume fractions of the suspension, a flocculated network of particles is formed throughout the sample whenever it is not being sheared, Under shear, however, this network is broken into smaller units of flocculated spheres which can withstand the shear forces (16). The size of the units which generally survive will be determined by a balance between the shear forces which tend to break the units down and the forces of attraction which hold the spheres together (17-19). The appropriate dimensionless group characterizing this process (balance of viscous and van der Waals forces) is $\eta_0 a^4 \gamma/A$, where A is the Hamaker constant.

Each flocculated unit is expected to rotate in the shear field, and it is likely that these units will tend to form layers as individual spheres do. As the shear stress increases, each rotating unit will ultimately behave as an individual sphere and, therefore, a flocculated suspension will show pseudoplastic flow, with the relative viscosity approaching a constant value (pseudo-Newtonian) at high shear rates. The viscosity-shear rate curve will also show a pseudo-Newtonian region at low and high shear rates, as with stable systems, although the values of the low and high shear rate viscosities (η_0 and η_∞) will of course depend on the extent of flocculation in the system and the volume fraction. It is also clear that such systems will show an apparent yield stress (Bingham yield value, τ_β), normally obtained by extrapolation of the linear portion of the τ-γ curve to $\gamma = 0$. Moreover, since the structural units in a flocculated system changes with change in shear, most flocculated suspensions show thixotropy. Once shear is initiated, some finite time is required to break the network of agglomerated units into smaller units which persist under the shear forces applied. As smaller units are formed, some of the liquid entrapped in the flocs is liberated, thereby reducing the effective volume fraction of the solid. This reduction in ϕ_{eff} is accompanied by a reduction in η_{eff} and this plays a major role in generating the thixotropy. Similar arguments may be invoked to account for the shear thinning behaviour of flocculated suspensions.

It is convenient to distinguish between two types of unstable systems, depending on the magnitude of the net attractive force. When this is relatively small, i.e. of the order of a few kT, the suspension is usually referred to as weakly flocculated. This is the case, for example, with suspensions that are flocculated in the secondary minimum or those with a "thin" adsorbed layer. Weak flocculation also occurs when a "free" (non-adsorbing) polymer is added to a stable suspension. However, in this case the attractive energy may reach several tens kT units. The second type of unstable suspensions are those where the net attraction is large, as for example the case of flocculation in the primary minimum (usually referred to as coagulated systems) or those flocculated by reduction of solvency (to worse than a θ-solvent) for the chains of a sterically stabilized suspension. The attractive energy in these cases is several hundred kT units. Below the rheology of these different systems will be described.

Weakly Flocculated Systems.

Weakly flocculated systems may be produced by the addition of free (non-adsorbing) polymer to a sterically stabilized suspension (20,21). Above a critical volume fraction of the free polymer, ϕ_p^+, weak flocculation occurs and this flocculation increases in magnitude with further increase in the free polymer concentration. Several rheological investigations of such systems have been carried out by Tadros and his Collaborators (22-24). The results showed an increase in the rhological paramers of the system (e.g. the extrapolated Bingham yield value or storgae modulus), when ϕ_p exceeds ϕ_p^+. The latter decreased with increase of the molecular weight of the polymer as predicted by theory.

It is possible, in principle, to relate the extrapolated Bingham yield stress, τ_β, to the energy required to separate the flocs into single units, E_{sep} (22,23),

$$\tau_\beta = \frac{3\,\phi_s\,n\,E_{sep}}{8\pi a^3} \tag{11}$$

where n is the average number of contacts per particle (the coordination number). The maximum value for n is 12 which corresponds to hexagonal or face-centered cubic lattice. This maximum value is highly unlikely for particle arrangement in a flocculated system. A more realistic value for n is 8, corresponding to random arrangement of particles in a floc, again assuming a compact structure. Again, this is probably unlikely and a more realistic value, corresponding to an open floc structure would be a significantly smaller value than 8.

In order to estimate E_{sep} from τ_β, a number of assumptions have to be made. Firstly, one has to assume that all of the particle-particle contacts are broken by shear. This may not always be realised, although viscosity measurements showed that the high shear value for a flocculated system is close to that of the latex before the addition of the free polymer. This implies that, under this high shear conditions, most of the particle-particle contacts are indeed broken. The second assumption that has to be made is the value to be assigned for n. As mentioned above, n becomes smaller the more open the floc structure is. It is, therefore, possible that n may not remain constant, depending on the extent of flocculation, which depends on the volume fraction of the free polymer as well as its molecular weight. For the sake of comparison, values of n varying between 4 and 12 were used and the results of the calculations were tabulated by Liang et al (24). The results were compared with theoretical values of the free energy of depletion, G_{dep} calculated from the Asakura and Oosawa (AO) (25) and Fleer, Scheutjens and Vincent (FSV) (26) theories on depletion flocculation. Close agreement between E_{sep} (assuming a value of n of 4) and G_{dep} based on Asakura and Oosawa's theory (25) although this should only be considered fortuitous.

Strongly Flocculated systems.

These can be exemplified by flocculation of sterically stabilised systems produced by reduction of the solvency of the medium for the stabiliser chain. For example, sterically stabilised latex dispersions containing polyethylene oxide) (PEO) can be flocculated by addition of electrolyte, e.g. Na_2SO_4 above a critical concentration (CFC). Alternatively, at a given electrolyte concentration the latex dispersion becomes flocculated above a critical temperature (CFT).

The flocculation of concentrated dispersions can be investigated using rheological measurements (27). This was illustrated using polystyrene latex dispersions containing grafted PEO chain, to which N_2SO_4 was added to reduce the solvency of the medium for the PEO chains. The yield value showed a rapid increase above a critical electrolyte concentration (>0.3 mol dm^{-3}). Similar results were obtained when the storage modulus was plotted versus electrolyte concentration.

The nature of the flocculated structure could be assessed using scaling laws using log-log plots of τ_β or G' versus ϕ_s at various Na_2SO_4

$$\tau_\beta = k \phi_s^m \tag{12}$$

$$G' = k' \phi_s^n \qquad (0.35 < \phi_s < 0.53) \tag{13}$$

where k and k' are constants and m and n are exponents. The values of m and n showed a sudden drop in m value from a value of ~ 31 to ~ 9.1 and of n from ~ 30 to ~ 12 as the Na_2SO_4 concentration is increased from 0.3 to 0.4 mol dm^{-3}. With further increase in Na_2SO_4 concentration from

182

0.4 to 0.5 mol dm^{-3}, m drops from 9.4 to 2.8, whereas n drops from 12 to 2.2. This low exponent is an indication that an open network floc structure with low fractal dimensions is formed.

The exponent of 2.8 or 2.2 at 0.5 mol dm^{-3} Na$_2$SO$_4$ is just in the range of the reported values in the literature. Many authors (27,28) have reported that the exponent for flocculated suspensions is in the range 2.0 - 4.5. However, the value of the exponent depends to some extent on the treatment a coagulated suspension has been subjected before the measurements were made.

Rheological results can also be applied to measure the critical flocculation temperature, CFT, of a suspension, by measuring the yild value or modulus as a function of temperature at a given electrolyte concentration. This was illustrated for the latex suspension in 0.2 mol dm^{-3} Na$_2$SO$_4$, which showed a rapid increase in the rhological parameters above 50°C. At 0.3 mol dm^{-3}, the rapid increase occured above 35°C, whereas at 0.4 mol dm^{-3} this increase occurred above 15°C.

Another example of strongly flocculated system is electrostatically stabilised polystyrene latex suspensions coagulated by addition of electrolyte, e.g. 0.2 mol dm^{-3} NaCl. In this case coagulation into the primary minimum occurs. The structure of such coagulated systems becomes partially broken down above a critical strain (deformation) that depends on the volume fraction of the suspension. Using scaling laws, one can obtain information on the structure of the flocculated system. Log-log plots of G' versus ϕ gave the following scaling equation,

$$G' = 1.98 \times 10^7 \phi^{6.0} \tag{14}$$

The high power in ϕ is indicative of a relatively compact flocculated structure. This is not surprising since the latex was coagulated at 0.2 mol dm^{-3} NaCl, which is not much higher than the CFC of the latex (~ 0.1 mol dm^{-3}).

REFERENCES

1. Rheology for Chemists, an Introduction, Goodwin, J.W. and Hughes, R.W., Royal Society of Chemistry Publication, Cambridge, U.K., 2000
2. Krieger, I.M. and Dougherty, M. Trans. Soc. Rheol.,1959, 3, 137.
3. Krieger,I.M., Adv. Colloid Interface Sci.,1972, 3, 111.
4. Bachelor,G.K., J. Fluid Mech., 1977, 83, 97.
5. Tadros, Th.F. and Hopkinson,A. Faraday Disc.Chem.Soc.,1990, 90,41.

6. Prestidge, C. and Tadros, Th.F., J. Colloid Interface Sci.,**1988**, 124, 660.
7. Liang, W., Tadros, Th.F. and Luckham, P.F., J. Colloid Interface Sci., **1992,** 153, 131.
8. Bromley, C., Colloids Surf., **1985**, 17, 1.
9. Costello, B.A. de L., Luckham, P.F., and Tadros, Th.F., Colloids Surf., **1988/1989,** 34, 301.
10. Luckham,P.F., Ansarifar, M.A.,Costello, B.A. de L. and Tadros, Th.F., Powder Technol.,**1991**, 65, 371.
11. Costello, B.A. de L., Luckham, P.F. and Tadros, Th.F., J. Colloid Interface Sci.,**1992**, 237.
12 Tadros, Th.F., Liang,W. Costello, B.A. de L. and Luckham, P.F., Colloids Surf.,**1993**, 79, 105.
13. White, L.R.,J. Colloid Interface Sci., **1983**, 95, 286.
14. de Gennes, P.G., Adv. Colloid Interface Sci.,**1987**, 27, 189.
15. Russel, W.B.,J. Rheol., **1980**, 24, 287.
16. Science and Technology of Polymer Colloids, Hoffman, R.L., Editors Poehlein, G.W., Ottewill, R.H., and Goodwin, J.W., Martinus Nijhoff Publishers, Boston, the Hague,**1983**, Vol.II,p. 570.
17 Firth, B.A. and Hunter, R.J. J. Colloid Interface Sci.,**1976**, 57, 248.
18. van de Ven, T.G.M. and Hunter, R.J. Rheol. Acta, **1976**, 16, 534.
19. Hunter, R.J. and Frayane,J., J. Colloid Interface Sci., **1980**, 76, 107.
20. Heath, D. and Tadros, Th.F.,Faraday Disc. Chem. Soc., **1983**, 76, 203
21. Prestidge, C. and Tadros, Th.F., Colloids Surf., **1988**, 31, 325.
22. Tadros, Th.F. and Zsednai, A., Colloids Surf., **1990**, 49, 103.
23. Liang, W.,Tadros, Th.F. and Luckham, P.F., J. Colloid Interface Sci., **1993**, 155, 156 (1993).
24. Liang, W., Tadros, Th.F. and Luckham, P.F., J. Colloid Interface Sci., **1993,** 160, 183.
25. Asakura, S. and Oosawa, F.,J. Polym. Sci., **1958**, 33, 183.
27. Fleer, G.J., Scheutjens, J.H.M.H. and Vincent, B., ACS Symposium Ser.,**1984**, 240, 245.
28. Liang, W.,Tadros, Th.F. and Luckham, P.F.,Langmuir,**1993**, 9, 2077.

Chapter 13

Polymer Adsorption and Conformation in Dispersion–Flocculation of Concentrated Suspensions

Zhonghua Pan[1], Ponisseril Somasundaran[1,*], and Laurence Senak[2]

[1]NSF Industry/University Cooperative Research Center for Advanced Studies in Novel Surfactants, Langmuir Center for Colloids and Interfaces, Columbia University, New York, NY 10027
[2]Research and Development, International Specialty Products, 1361 Alps Road, Wayne, NJ 07470

Performance of polymers as stabilizers or flloccultants for controlling the properties of <u>concentrated</u> suspensions depends upon both the extent of the polymer adsorption as well as conformation at solid-liquid interfaces. Solids loading, polymer molecular weight and fractionation, and dissolved solid species, among others, are important factors that control the polymer adsorption and conformation and the system behavior. In this study, polymer molecular weight fractionation and adsorption changes were monitored using stepwise adsorption tests with techniques such as TOC (total organic carbon) and GPC/LS (gel-permeation chromatography /light scattering). Fluorescence and ESR (electron spin resonance) spectroscopy were used to monitor the conformation/orientation of the polymers adsorbed on the solid particles. Dissolved aluminum species in the residual solutions was monitored using ICP (inductively coupled plasma) atomic emission. Zeta potential measurements of the suspensions were made using a Zeta-Meter Model D to investigate effects of the dissolved aluminum species on the system behavior. The adsorption of PAA on alumina did not change significantly with increase in solids loading from 2 vol. % to 15 vol. % as a whole. However, there is large scattering of data at low solids loading (2 vol. %) and fixed initial high PAA concentration as found in our previous study which showed a "greater" adsorption density at low solids loading with a marked decrease in adsorption with increase in solids loading in the same range. Interestingly, the polymer adsorption behavior at high solids loading is different from that at low solids loading when the polymer sample contains impurities such as dioxane. Under the test conditions, smaller polymer molecules were found to preferentially adsorb first at the interfaces. Polymer molecular weight fractionation due to such preferential adsorption is more evident at low solids loading than at high solids loading, suggesting an increasing effect of particle-particle interaction on the polymer diffusion. Concentration of the dissolved aluminum species in the residual solutions increased significantly as the system becomes denser. Dissolved aluminum species was found to affect the conformation of PAA in the solution: PAA molecules become more coiled in the presence of the species at pH<7 even though the concentration of the species is low. Zeta potential of PAA-alumina suspensions with additions of dissolved aluminum species was markedly higher than that of PAA-alumina systems at pH 4 to pH 10, suggesting that the alumina particle surface becomes more positive in this pH range due to increasing complexation of PAA with the dissolved alumina species. Our earlier fluorescence and ESR results had suggested that the adsorbed PAA molecules tend to stretch out or dangle more into the solution as the system becomes denser. It is clear that conformation of the adsorbed polymer is an important parameter for controlling flocculation/dispersion behavior of concentrated suspensions.

Introduction

The colloidal dispersions encountered in many industrial processes invariably involve concentrated or dense suspensions where the dispersed phase can be as large as 50 % or more by volume. Control of the state of concentrated suspensions for efficient performance is a critical and difficult issue to be dealt with in the industry. However, extensive studies on both theoretical and experimental aspects of polymer adsorption have been conducted almost exclusively with dilute suspensions (1- 4), and hence there is very little known on the polymer behavior at interfaces in concentrated suspensions. While there have been many studies on the behavior of concentrated suspensions (5-32) themselves, there have been virtually none dealing with the important aspect of polymer conformation/orientation as they relate to suspension properties. This is mainly due to the inherent difficulties, such as lack of suitable techniques and adequate equipment, as well as poor theories for studying polymer adsorption and conformation *in situ* in such dense systems. Recently, several spectroscopic techniques (*33-47*), such as fluorescence, electron spin resonance (ESR), nuclear magnetic resonance (NMR) and Fourier-transform infrared (FTIR) spectroscopy, as well as small angle neutron scattering (SANS), have been used to help determine the fraction of polymer segments bound on surfaces as well as the distribution of segments in the vicinity of the surface. It is to be noted here that all these studies apply to *dilute* suspensions.

The features of concentrated suspensions differ from those of their dilute counterparts in many ways. For example, rheological behavior of concentrated silica suspensions showed a sharp rise in relative viscosity for solids fractions above 25 % (*48*). This is mainly due to the increase in multiparticle interactions in concentrated systems since such interactions are strongly dependent on the separation distance between the suspended particles in them. Multiparticle interactions may also affect polymer adsorption and conformation behavior at solid-liquid interfaces in such concentrated systems. On the other hand, changes in dissolved solid species and their concentration due to changes in pH and solids contents and possible complexation of these species with polymers can also have measurable effects on the polymer behavior. This article will present the results of our recent investigation on solids loading, polymer molecular weight and fractionation, as well as dissolved aluminum species on the polymer adsorption and conformation and zeta potential of suspensions, along with a brief review of our previous observations in monitoring of the polymer conformation in concentrated systems.

Materials and Methods

Materials

The results discussed here were obtained from tests with two systems: polyacrylic acid (PAA) - alumina and polyethylene oxide (PEO) - silica. Both these systems are widely used in the industry.

The alumina used was AKP-50 powder from Sumitomo Chemical Inc. Purity of the alumina powder was specified by the manufacturer to be 99.9%. The particles, with a density of 3.97 g cm^{-3}, were nonporous as indicated by scanning electro microscopy. The average particle diameter was 0.21 µm and the specific surface area as determined by nitrogen B.E.T. adsorption to be 10.9 m^2 g^{-1}. The silica was obtained from Geltech Inc. and had a nominal particle size of 1 μm and a specific surface area of $4.2m^2 g^{-1}$. The silica particles, of a density of 2.1 g cm^{-3}, were nonporous.

Unlabeled PAA, with a specified molecular weight of 90,000 and polydispersity of 2.98 ~ 3.7, was purchased from Polysciences Inc. Pyrene-labeled PAA, with a pyrene content of roughly 0.74 percent and a molecular weight of 76,000, was synthesized by National Chemical Laboratory, India. Nitroxyl-labeled PAA with a molecular weight of 95,000 and a labeling ratio of 1:100 (molar basis) was prepared by Nalco Chemical Company. Unlabeled PEO, of average molecular weight between 6,000 and 7,500, was obtained from Polysciences Inc. Pyrene-labeled PEO, with a 1 percent pyrene content and a molecular weight of 7,500, was synthesized by Kumar and Aguilar in the Department of Chemistry at the University of Florida.

A 0.03 M sodium nitrate solution was used for ionic strength control in preparing the polymer solutions and solid suspensions.

Methods

Polymer Adsorption

Polymer adsorption on solids was determined by monitoring depletion of the polymer from the solutions after equilibration with solids using total organic carbon (TOC) analysis.

Polymer Conformation

The conformation of pyrene-labeled PAA and PEO was determined by analyzing the fluorescence spectra of the suspensions collected using an LS-1 fluorescence spectrometer (Photon Technology International Inc.). Mixtures of 3 % pyrene-labeled PAA with 97 % unlabeled PAA in PAA-alumina systems, and 2 % pyrene-labeled PEO with 98 % unlabeled PEO in PEO-silica systems, were used to minimize any perturbative effects of the pyrene moieties. The effect of pyrene hydrophobicity on the solution behavior of pyrene-containing systems has been established (49,50). More recently, the pyrene label was found to affect significantly the stability of silica suspensions containing pyrene-labeled polymers (51). However, a low pyrene content, for example, 3 % pyrene-labeled PEO in a silica-PEO system was found not to have significant effects on the flocculation behavior of the system compared with such a system without any pyrene-labeled PEO (51). This suggests that at a low pyrene content in the labeled polymer and a low ratio of pyrene-labeled polymer in the mixtures, the behavior of the labeled polymer molecules is similar to that of the unlabeled ones as a whole.

The coiling index of pyrene-labeled PAA and PEO was used as an *in-situ* measure of the polymer conformation. This parameter is determined from the ratio of fluorescence emission intensity of the excimer peak at around 475~480 nm (Ie) to that of the monomer peak at around 375 nm (Im) in the pyrene fluorescence spectrum (*43*). A higher ratio, therefore, indicates presence of much pyrene excimers and hence a more coiled conformation of the polymer, while a lower ratio suggests a more extended conformation (Figure 1).

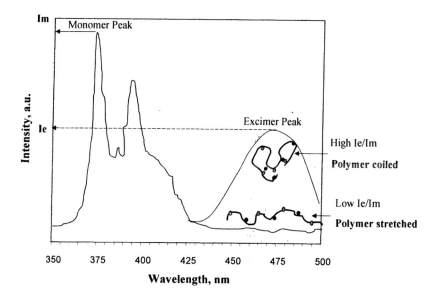

Figure 1 Schematic representation of pyrene emission spectrum and its use in the determination of polymer conformation

Molecular species with a free electron possess intrinsic angular momentum (spin), which in an external magnetic field undergoes Zeeman splitting. In electron spin resonance (ESR) spectroscopy, changes in the spin characteristics of a paramagnetic probe in an external magnetic field are monitored (*52*) in the form of absorption of microwave radiation. The line positions and splitting in ESR spectra depend on the direction of external magnetic field relative to the molecular axis. This phenomenon is called anisotropy. Information on the microviscosity and micropolarity of a spin probe environment can be obtained from changes in the ESR spectral line shape of this probe. The former measure is obtained from the spectrum in terms of a calculated rotational correlation time (τ), the time required for spin label to rotate through an angle of one radian, while the latter is obtained from the measured hyperfine splitting constant (*44*). Spectral anisotropy of nitroxide spin label under various conditions of motion is illustrated in Figure 2. The rotational mobility of the nitroxide spin label under different conditions can be estimated from these ESR spectra (*53*). Specially, the mobility of nitroxide label in slow tumbling ESR spectra, which is typically the case for adsorbed polymeric nitroxide probe in concentrated particulate systems, can be obtained by calculating the rotational correlation time (*54*). The mobility of nitroxide label gives information on the conformation and mobility of the labeled polymer segments.

In our experiments, mixtures of 5 % nitroxyl-labeled PAA (label molar ratio 1:100) with 95 % unlabeled PAA were employed to monitor the mobility of adsorbed polymer. ESR spectra were collected using a Model 8300A X-Band spectrometer (Micro-Now Instrument Company) equipped with a 9 GHz microwave frequency generator.

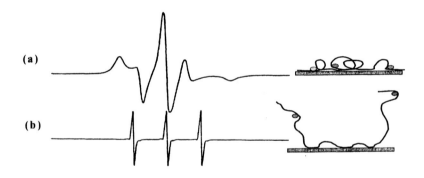

Figure 2. Representative ESR spectra for a nitroxide-labeloed polymer and the corresponding suggested orientation/mobility.
(a) Anisotropic spectrum: nitroxide probes rotate lowly, suggesting a <u>restricted</u> orientation (e. g., when the polymer is coiled or extended at the solid surface);
(b) Isotropic spectrum: nitroxide probes freely rotate in the solution, suggesting a non-restricted environment (e. g., when the polymer dangles into the solution).
The rotational mobility of nitroxide label can be qualified by the rotational correlation time (τ):$\tau_a > \tau_b$

Polymer Molecular Weight Fractionation

Polymer molecular weight fractionation was analyzed using GPC/LS (gel-permeation chromatography/light scattering) in order to obtain absolute molecular weight distribution. The instrumentation employed for this work included a Wyatt Technologies DAWN DSP multi-angle light scattering photometer set in tandem to a Waters gel permeation chromatography system, including refractive index detection (The Waters system included a model 590 solvent delivery system, a 410 refractive index detector, and a WISP 717 auto-sampler, all thermostated to 30 °C). A single Shodex SB-80MHQ linear GPC column was employed for this work. The mobile phase used was a pH-7 aqueous buffer that was 0.1 molar TRIS and 0.2 molar lithium nitrate (brought to pH with nitric acid). 100 μl of 0.2 % w/v of PAA sample concentration was introduced to the chromatography system. A dn/dc value of 0.250 ml/g was applied to these experiments and all data was analyzed using Wyatt ASTRA software version 4.50.

Stepwise adsorption tests were set up (Figure 3) to investigate such fractionation and its effects on the polymer adsorption and conformation in polydispersed polymer-particle systems.

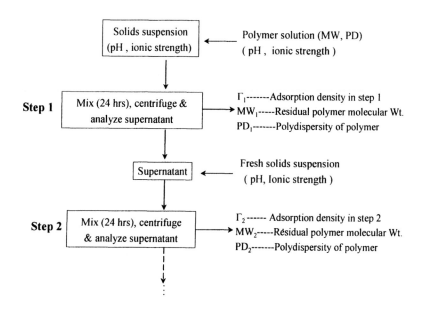

Figure 3   Flow sheet of Stepwise Adsorption (@ pH 4 and I=0.03 M NaNO$_3$)

Aluminum Concentration

Aluminum concentration in the residual solutions was determined by Perkin-Elmer ICP/6500 emission spectroscopy.

Zeta Potential of Suspensions

Zeta potentials were measured using a Zeta meter Model D.

Results and Discussion

Polymer Adsorption vs. Solids Loadings and Impurity

Effect of solids loading on the polymer adsorption and conformation at the solid-liquid interfaces in concentrated suspensions is a key issue to be dealt with in developing mechanisms of flocculation/dispersion of such suspensions. Our previous adsorption test results have shown that solids loading has a marked effect on the adsorption density of PAA on alumina and PEO on silica (55,56): increase in solids loading was found to result in a decrease in the adsorption of the polymers. This effect was especially evident in the low solids loading range (less than 10 vol. % solids) at fixed initial high polymer concentration. One possible explanation for the "greater" adsorption density under these conditions has to do with precipitation/multilayer adsorption of the polymer under such conditions. Addition of small numbers of particles to a solution at a high polymer concentration may in fact result in the precipitation or multilayer adsorption of the polymer on the solids and thus a "greater" adsorption density, while solids surface area is no longer limited at high solids loading and the adsorption density approaches the "real" value. Additional experiments have been done recently to test these possibilities. The results obtained from these experiments are shown in Figure 4. It can be seen from Figure 4 that the adsorption density stayed within the experimental error with the increase in solids loading. Scattering of data is however rather large under these conditions. In addition to possible precipitation/multiplayer that results in a "greater" adsorption density, the "lower" or "greater" adsorption density *at low solids loading and fixed initial high polymer concentration* can also result from the relative error in measurements and calculation of the residual polymer concentration since even a small error could result in significant differences in the adsorption density.

Interestingly, it was found in our experiments that the polymer adsorption behavior at high solids loading is different from that at low solids loading when the polymer sample contains impurities such as dioxane (Figure 5)[*]. It can be seen that at low solids loading (1.3 vol. %), the isotherm of dioxane-containing PAA adsorption is similar to that of PAA without such an impurity. However, at high solids loading (15 vol. %), the isotherm is very different from that of PAA without dioxane. There still is significant polymer in the residual solution even though the initial polymer concentration is low and the solid surface is not saturated by the polymer. Reasons for this effect are not clear at present.

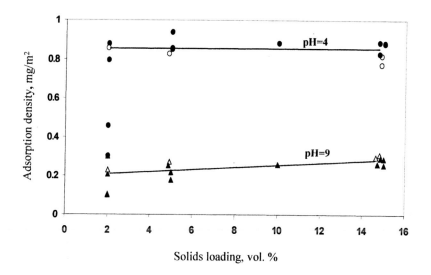

Figure 4 Polyacrylic acid adsorption vs. solids loading (PAA Mw=90,000; I=0.03 M NaNO$_3$, AKP-50 alumina)
Closed symbol: PAA adsorption at fixed initial high concentrations; Open symbol: PAA adsorption at reduced initial concentrations.

Polymer Molecular Weight Fractionation in PAA-Alumina Systems

It is well known that polymer adsorption on solid increases with increase in its molecular weight. For example, in PAA-alumina systems, the adsorption

---

[*] In collaboration with Dr. B. Pethica

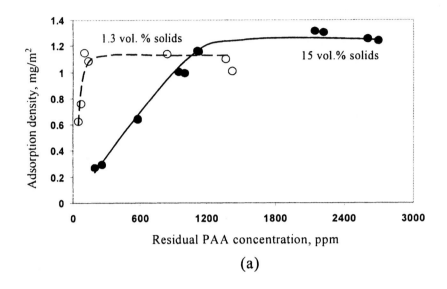

(a)

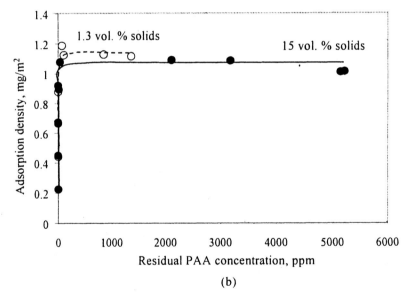

(b)

Figure 5 Polyacrylic acid adsorption on alumina
(pH=4, I=0.03 M NaNO$_3$, AKP-50 alumina)

(a) Dioxane-containing PAA (PAA Mw=148,200, from Polymer Sources Inc.).
(b) PAA without dioxane (PAA Mw =90,000, from Polyscience, Inc.)

density of a 150,000 molecular weight PAA is much higher than that of a 2000 PAA (Figure 6). Such changes in adsorption due to different molecular weight (different chain of polymers) could also lead to changes in the polymer conformation. In this study, stepwise adsorption tests (see Figure 3) were done to investigate possible polymer molecular weight fractionation in concentrated alumina-polydispersed PAA systems since such fractionation could also result in changes in adsorption, which in turn could lead to changes in the polymer conformation.

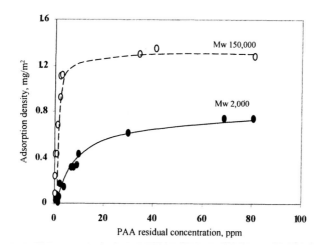

Figure 6. PAA molecular weight effect on adsorption
(AKP-50 alumina, 5 wt.% solids loading, pH 4, I=0.03 M NaCl)

Results from these tests showed that both the polymer molecular weight of the solutions and the adsorption increase with increase in adsorption steps at both high and low solids loadings (Figures 7). Gel-permeation chromatographic/light scattering analysis clearly revealed a shift of the polymer molecular weight in the solutions to higher values with increase in adsorption steps (Figures 8, 9), suggesting that smaller polymer molecules are adsorbed preferentially at the interface in the earlier stages. Compared with evident fractionation at low solids loading (1.3 vol.%), there is less fractionation in the first adsorption step and a marked fractionation in the second step at high solids loading (15 vol.%), implying that it is more difficult for the polymer species to fractionate in concentrated particulate systems possibly due to difficulties for the polymer species to diffuse to the surface preferentially in such systems.

These findings about polymer molecular weight fractionation and the difference in the extent of such a fractionation between dilute and concentrated systems have important implications in determining the mechanisms involved in changes in the polymer conformation. Monitoring of such changes due to molecular weight and fractionation by designing mixed labeled monodispersed polymers of different molecular weights and with different order of addition of those polymers is necessary in all investigation of polymer adsorption.

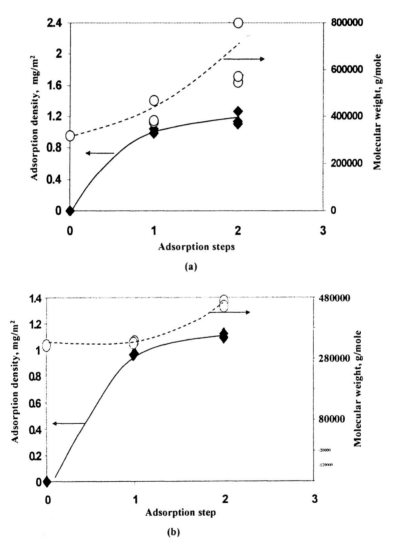

Figure 7 PAA adsorption and residual PAA molecular weight as a function of adsorption step

(Polyacrylic acid Mw ~ 90,000; AKP-50 alumina; pH=4; I=0.03 M NaNO$_3$)

(a) 1.3 vol. % solids loading. (b) 15 vol. % solids loading.

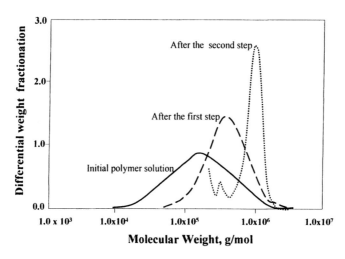

Figure 8 PAA molecular weight distribution of the solutions as a function of adsorption step at low solids loading (Polyacrylic acid Mw=90,000, AKP-50 alumina, **1.3 vol. % solids loading**, pH=4, I=0.03 M NaNO$_3$)

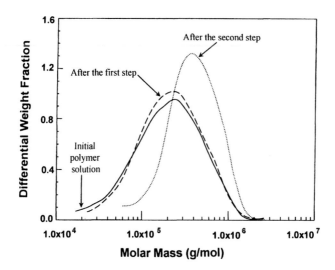

Figure 9 PAA molecular weight distribution of the solutions as a function of adsorption step at high solids loading (PAA Mw~90,000; AKP-50 alumina ,**15 vol.% solids loading**; pH=4; I=0.03 M NaNO$_3$)

Conformation and Mobility of Adsorbed Polymers

Our fluorescence work on PAA-alumina systems and PEO-silica systems showed that the coiling index (Ie/Im) of the pyrene-labeled polymers decreased with increase in solids loading, suggesting that the adsorbed polymer molecules tend to become extended as the system becomes denser (55,56). ESR spectroscopy of absorbed nitroxide-labeled PAA on alumina showed that the portion of non-restricted (free rotational) components in the spectra increased with increase in solids loading, suggesting that the portion of "loops" and/or "tails" of the adsorbed polymer chains increased with increase in solids loading. The average rotational correlation time calculated based on these ESR spectra decreases with increase in solids loading, implying that the adsorbed polymer segments tend to become less restricted with increase in solids loading, which is the case when the adsorbed polymer segments stretch out or dangle into the solution (55, 56).

Dissolved Alumina Species and Their Effects on the PAA Conformation and Zeta Potential of Suspensions

When alumina is brought into contact with water, its dissolution is followed by pH-dependent hydrolysis and complexation of the dissolved species with polymers in solution, which in turn can affect polymer adsorption and conformation, as well as the stability of resultant suspensions. Therefore it is necessary to investigate aluminum concentration in residual solutions and its effects.

Aluminum concentration of the residual solutions was found to increase significantly with increase in solids loading (Figure 10). Effect of such dissolved aluminum species on the conformation of polyacrylic acid (PAA) is illustrated in Figure 11. Supernatants containing dissolved aluminum species were obtained from 5 wt.% (~ 1.3 vol.%) alumina suspensions. It can be seen that the coiling index (Ie/Im) for PAA in the dissolved aluminum species-PAA solution is measurably higher than that for PAA in the PAA solution alone at pH < 7, which suggests that the PAA molecules do become more coiled in the presence of the dissolved aluminum species at low pH values. At pH 5, for example, PAA could complex with Al^{3+}, $Al(OH)^{2+}$ and $Al(OH_2)^+$ due to the electrostatic interactions between these positively charged species and the COO^- groups on the polymer, leading to a lowering of the electrostatic repulsion and more coiling of the polymer chain; hence the higher Ie/Im ratio for PAA/dissolved species systems. With increase in pH, even though the polymer is still ionized, because of the decrease in the concentration of the positive charged species, the extent of interaction between the polymer and the species is reduced. Consequently due to the increase in the intra-polymer and polymer/$Al(OH)_4^-$ repulsion, PAA retains its stretched conformation (lower Ie/Im ratios).

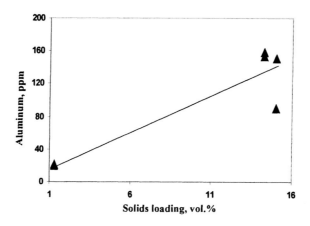

Figure 10 Aluminum concentration of the residual solutions as a function of solids loading (AKP-50 alumina; pH=4; I=0.03 M NaNO$_3$)

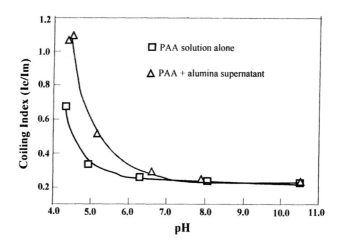

Figure 11 Coiling index (Ie/Im ratio) for Polyacrylic acid solutions alone And Polyacrylic acid-alumina supernatants as a function of pH (PAA Mw=90,000, PAA concentration=100 ppm; I=0.03 M NaCl)

Effects of dissolved aluminum species on zeta potential of PAA-alumina suspensions are shown in Figure 12. In this study, PAA solutions were prepared using triply distilled water and alumina supernatants that were obtained from 5 wt.% alumina suspensions, respectively. The PAA solution was then added to the alumina suspensions. Zeta potential is seen to be markedly higher for alumina supernatant-PAA/alumina suspensions than that for PAA/alumina suspensions at pH 4 to pH 10, which implies that the alumina particle surface becomes more positive in the former system possibly due to increasing complexation of PAA with the dissolved aluminum species from both the alumina suspension itself and the introduction of alumina supernatants.

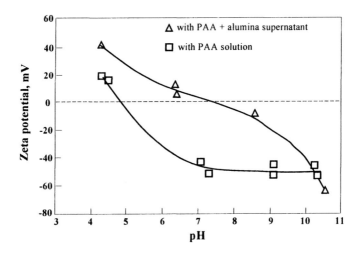

Figure 12 Zeta potential of alumina with PAA solution and with PAA solution -alumina supernatant as a function of pH (PAA Mw=90,000, PAA concentration=100 ppm; solids loading=5 wt. %; I=0.03 NaCl)

These results have enabled us to narrow the list of possible factors controlling the polymer behavior in concentrated systems. It is clear that monodispersed polymer samples with fluorescence and spin labels are required to monitor the polymer conformation and to ascertain the mechanisms involved in changes in the polymer conformation. Further controlled experiments on the conformational studies are also needed to monitor and minimize perturbative effects of these labels on the systems.

Summary

1. Under the test conditions described in this· article, polyacrylic acid adsorption on alumina did not show significant changes with increase in solids loading. Interestingly, the polymer adsorption behavior at high solids loading is different from that at low solids loading when the polymer sample contains impurities such as dioxane. Our earlier fluorescence and ESR results implied that the adsorbed polyacrylic acid and polyethylene oxide molecules tend to stretch out or dangle more into the solution as the system becomes denser.

2. Marked molecular weight fractionation was observed at both low and high solids loadings in the stepwise adsorption tests, suggesting that smaller molecules are adsorbed preferentially at the solid-liquid interfaces in the early stage of adsorption. However, such fractionation in the system at high solids loading occurred only in the second step, indicating that it is more difficult for the polymer to diffuse to the surface preferentially in such systems. Polymer molecular weight fractionation leads to changes in its adsorption. This could in turn result in changes in the polymer conformation.

3. Concentration of dissolved aluminum species increased significantly with increase in solids loading. These dissolved aluminum species showed measurable effects on polyacrylic acid conformation as well as system behavior such as zeta potential and flocculation/dispersion.

Acknowledgments

The authors acknowledge the support of the Engineering Research Center (ERC) for Particle Science and Technology at the University of Florida (National Science Foundation (NSF) grant # EEC-94-02989), the Industry/University Cooperative Research Center (IUCRC) for Advanced Studies in Novel Surfactants at Columbia University (NSF grant # EEC-98-04168), the U.S Department of Energy, and the Industrial Partners of the IUCRC: International Specialty Products, Dispersion Technology Inc., Akzo Nobel, Hercules, Uniliver, and Sun Chemical, Inc. The authors also acknowledge Indian National Chemical Laboratory for synthesizing pyrene-labeled polyacrylic acid, Kumar and Aguilar in the Department of Chemistry at the University of Florida for synthesizing pyrene-labeled polyethylene oxide, Nalco Chemical Company for synthesizing nitroxyl-labeled polyacrylic acid, and Dr. Berislav Markovic in the International Specialty Products for his help.

References

1. Sato, T.; Ruch, R. In *Surfactant Science Series*; Sato, T.; Ruch, R., Eds.; Marcel Dekker: New York, N Y, 1980; Vol. 9, pp 65-120.
2. Fleer, G. J.; Cohen Stuart, M. A.; Scheutjens, J. M. H. M.; Cosgrove, T.; Vincent, B. In *Polymers at Interfaces*; Fleer, G. J.; Cohen Stuart, M. A.; Scheutjens, J. M. H. M., Eds.; Chapman & Hall: London, 1993.
3. Fleer, G. J.; Scheutjens, J. M. H. M. In *Surfactant Science Series*; Dobias, B., Ed.; Marcel Dekker: New York, N Y, 1993; Vol. 47, Chapter 5.
4. Markovic, B. Ph. D. thesis, University of Zagreb, Zagreb, Croatia, 1996.
5. Chander, S. *Colloids Surf A.* **1998,** *133,* 143-150.
6. SenGupta, A. K.; Papadopoulos, K. D. *J. Colloid Interface Sci.* **1998,** *203,* 345-353.
7. Ohshima, H. *J. Colloid Interface Sci.* **1997,** *188,* 481-485
8. Evanko, C. R.; Dzombak, D. A.; Novak, Jr., J. W. *Colloids and Surf A.* **1996,** *110,* 219-233.
9. Huynh, L.; Jenkins, P.; Ralston, J. *Int. J. Min. Proc.* **2000,** *59,* 305-325.
10. Bergström, L. *Colloids Surf A.* **1998,** *133,* 151-155.
11. Anklekar, R. M.; Borkar, S. A.; Bhattacharjee, S.; Page, C. H.; Chatterjee, A. K. *Colloids Surf A.* **1998,** *133,* 41- 47.
12. Dukhin, A. S.; Shilov, V. N.; Ohshima, H.; Goetz, P. J. *Langmuir.* **1999,** *15,* 6692-6706.
13. Carrique, F.; Arroyo, F. J.; Delgado, A. V. *J. Colloid Interface Sci.* **2001,** *243,* 351-361.
14. Ottewill, R.H. In *Science and Technology of polymer Colloids;* Ottewill, R.H.; Poehlein, G. W.; Goodwin, J. W., Eds.; Martinus Nijhoff: Boston, MA, 1983; Vol. II, pp 503-522.
15. R.H. Ottewill, in: R.H. Ottewill, J.W. Goodwin (Eds.), Concentrated Dispersions, 1. Fundamental Considerations in Science and Technology of Polymer Colloids, vol. II, Martinus Nijhoff, Boston, 1983.
16. Tadros, Th.F. *Colloids Surf A.* **1986,** *18,* 137.
17. Shih, W.; Kim, S. I.; Shih, W.Y.; Schilling, C. H.; Aksay, I. A. *Mater. Res. Soc. Symp. Proc.* 1990, *180,* 167.
18. Frith, W. J.; Mewis, J.; Strivens, T. A. *Powder Technol.* **1987,** *51,* 27.
19. Mondy, L. A.; Graham, A. L. In *Better Materials Through Chemistry*; Zelinnski, B.J.J.; Brinker, C.J.; Clark, D.E.; Ulrich, D. R., Eds.; *Mater. Res. Soc. Symp. Proc.* 1990, *180,* 173.
20. Johnson, S. B.; Franks, G. V.; Scales, P. J.; Boger, D.V.; Healy, T. W. *Int. J. Min. Proc.* **2000,** *58,* 267.

21. Spicer, P.T.; Pratsinis, S. E.; Willemse, A.W.; Merkus, H. G.; Scarlett, B. *Part. Part. Syst. Char.* **1999,** *16,* 201.
22. Yziquel, F.; Carreau, P. J.; Moan, M.; Tanguy, P. A. *J. Non-Newtonian Fl. Mech.* **1999,** *86,* 133.
23. Ohshima, H. *Colloids Surf A.* **1999,** *153,* 477.
24. Lemke, T.; Bagusat, F.; Koehnke, K.; Husemann, K.; Moegel, H. *Colloids Surf A.* **1999,** *150,* 283.
25. Johnson, S. B.; Russell, A. S. *Colloids Surf A.* **1998,** *141,* 119.
26. Scales, P. J.; Johnson, S. B.; Healy, T. W.; Kapur, P. C. *AIChE J.* **1998,** *44,* 538.
27. Tsui, O. K. C.; Mochrie, S. G. J. *Phys. Rev. E.* **1998,** *57,* 2030.
28. Hoffman, R. L. *J. Rheol.* **1998,** *42,* 111.
29. Dratler, D. I.; Schowalter, W. R.; Hoffman, R. L. *J. Fluid Mech.* **1997,** *353,* 1.
30. Overbeck, E.; Sinn, C.; Palberg, T.; Schaetzel, K. *Colloids Surf A.* **1997,** *122,* 83.
31. Koh, C. J.; Hookham, P.; Leal, L. G. *J. Fluid Mech.* **1994,** *266,* 1.
32. Zupancic, A.; Lapasin, R.; Kristoffersson, A. *J. Eur. Cer. Soc.* **1998,** *18,* 467.
33. Yu, X.; Somasundaran, P. *J. Colloid Interface Sci.* **1996,** *177,* 283-287
34. Tjipangandjara, K. F.; Somasundaran, P. *Colloids Surf.* **1991,** *55,* 245-255.
35. Somasundaran, P.; Tjipangandjara, K. F.; Maltesh, C. In *Solid/Liquid Separation: Waste Management and Productivity Enhancement;* Muralidhara, H. S., Ed.; Battelle Press: Columbus, OH, 1989; pp 325-342.
36. Fan, A.; Somasundaran, P.; Turro, N. J. *Colloids Surf A.* **1999,** *146,* 397-403.
37. Somasundaran, P.; Yu, X.; Krishnakumar, S. *Colloids Surf A.* **1998,** *133,* 125-133.
38. Sivadasan, K.; Somasundaran, P. *Colloids Surf.* **1990,** *49,* 229-239.
39. Krishnakumar, S.; Somasundaran, P. *Colloids Surf A.* **1996,** *117,* 37- 44.
40. Krishnakumar, S. Somasundaran, P. *J. Colloid Interface Sci.* **1994,** *162,* 425-430.
41. Malbrel, C. A.; Somasundaran, P. *Langmuir,* **1992,** *8,* 1285-1290.
42. Somasundaran, P.; Turro, N. J.; Chandar, P. *Colloids Surf.* **1986,** *20,* 145-150.
43. Chandar, P.; Somasundaran, P.; Turro, N. J.; Waterman, K. C. *Langmuir,* **1987,** *3,* 298-300.
44. Chandar, P.; Somasundaran, P.; Waterman, K. C.; Turro, N. J. *J. Phys Chem.* **1987,** *91,* 150-154.
45. Somasundaran, P.; Kunjappu, J. T. *Colloids Surf.* **1989,** *37,* 245-268.

46. Somasundaran, P.; Chandar, P.; Chari, K. *Colloids Surf.* **1983**, *8*, 121-136.

47. Somasundaran, P.; Kunjappu, J. T.; Kumar, C. V.; Turro, N. J.; Barton, J. K. *Langmuir,* **1989**, *5*, 215-218.

48. Zaman, A. A.; Moudgil, B. M.; Fricke, A. L.; El-Shall, H. *J. Rheol.* **1996**, *40*, 6, 1191-1210,

49. Char, K.; Frank, C. W.; Gast, A. P.; Tang, W. T. *Macromolecules.* **1989**, *22,*1255.

50. Quina, F.; Abuin, E.; Lissi, E. *Macromolecules.* **1990**, *23*, 5173.

51. Campbell, A.; Somasundaran, P. *J. Colloid Interface Sci.* **2000**, *229*, 257-260.

52. Wertz, J. H.; Bolton, J. R. *Electron Spin Resonance*; Chapman and Hall: New York, NY, 1986.

53. Knowles, P. F.; Marsh, D.; Rattle, H. W. E. *Magnetic Resonance of Biomolecules: An Intrduction to the Theory and Practice of NMR and ESR in Biological Systems*; John Wiley and Sons: New York, NY, 1976.

54. Freed, J. In *Spin Labeling: Theory and Applications;* Berlinger L. J., Ed.; Academic Press: New York, NY, 1976; pp.72-85

55. Pan, Z.; Campbell, A.; Somasundaran, P. *Colloids Surf A.* **2001**, *191,*71-78

56. Campbell A.; Pan, Z.; Somasundaran P. In *Polymers in Particulate Systems: Properties and Applications;* Hackley V. A.; Somasundaran P.; Lewis J. A. Eds.; Marcel Dekker : New York, N Y, 2002; Chapter 5, pp135-156.

Chapter 14

Conformation and Aggregation Properties of Associative Terpolymer in Aqueous System and Correlation with Rheological Behavior

Anjing Lou, Ning Wu, and Richard Durand, Jr.

Daniel J. Carlick Technical Center, Sun Chemical Corporation, 631 Central Avenue, Carlstadt, NJ 07072

Abstract

The conformation and the aggregation properties of the HEURASE thickener, UCAR POLYPHOBE 106HE, at different pH and concentration in aqueous solutions have been studied by using Fluorescence and Dynamic Light Scattering techniques. It was found that the polymer possesses a coiled structure at low pH and stretched at high pH, which corresponds to the particle size variation. Since the pH goes down with the increase of the concentration, at about 0.02 wt% corresponding to pH 5.6 the molecules reach a minimum size. The polymer is completely stretched when the solution pH is above 7.8 and the molecular aggregates are formed when the concentration is high enough. However, the aggregate size is limited before cross-linking. The sharp increase of the solution viscosity at pH 7.8 is believed to be due to the formation of cross-link network.

Introduction

Hydrophobically modified water-soluble polymers, better known as associative thickeners, have achieved commercial acceptance in controlling the rheological properties in products such as paints, inks and paper coatings. Among these thickeners there are three main types of polymer backbones: ethylene oxide-urethane block copolymers, carboxylated polyacrylates and cellulose derivatives. The first type is also known as HEUR (hydrophobically modified ethoxylated urethane) thickeners (1). These associative thickeners usually provide the expectation of high film build and excellent leveling characteristics. Other important properties provided are reduced spatter and a non-flocculative thickening mechanism (2) resulting in increased gloss and hiding (3). Because of these advantageous properties, associative thickeners have drawn considerable attention from both industrial and academic professionals, especially for HEURASE (3-6) (hydrophobically modified ethoxylated urethane alkali-swellable/soluble emulsions). These polymers provide rheology comparable to that of an HEUR with the handling convenience of an ASE (alkali-swellable/soluble emulsion). These thickeners are ter-polymers produced by the emulsion polymerization of a carboxyl-functional monomer, a relatively water-insoluble monomer and a hydrophobe-terminated urethane-functional exthoxylated macromonomer (5,7,8). The general structure of these thickeners is shown in figure 1 (4,8), in which R = $C_{33}H_{51}O_2$.

Most of the published studies on associative thickeners have been focused on their rheology properties by determining the variation of the viscosity with pH or concentration and interpreted by proposing different models (1,9). It is generally agreed that the hydrophobic modifications of thickeners aggregate in an aqueous medium, leading to the formation of large networks of the polymers. A qualitative model describing such networks was first proposed by Bieleman et al (9). Further it is assumed that the surfactants interact with the hydrophobic aggregates formed by the associative thickeners, resulting in some form of mixed surfactant-thickener aggregates that affects the network structure and thus the rheology (1).

While the formation of a network is an important mechanism that would be helpful in understanding the system rheology performance, equally important is the conformation and orientation of the polymers in solution and at solid/liquid interfaces, especially for the systems such as various printing inks and paints, in which polymers are introduced to control the dispersity and stability as well as viscosity. It is well known that particle dispersion/aggregation is highly dependent on the polymer conformation in solution and orientation at particle surfaces (10).

In this work, we investigated the molecular conformation and particle size of the terpolymer 106 HE, one of the HEURASE products from Union

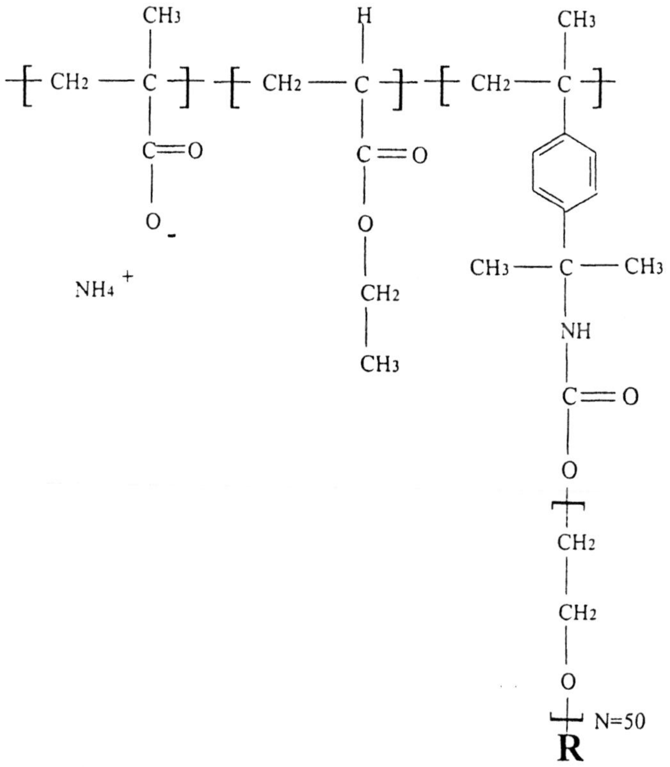

Figure 1. Alkali-soluble thickener structure

Carbide Corp., at different pH and concentrations in aqueous systems by using pyrene fluorescence and dynamic light scattering techniques. The results were then compared with the viscosity measurements.

Experimental

Materials:

The HEURASE thickener emulsion, 106HE, was obtained from Union Carbide Corp. The concentration was based on the solid material percentage. The water we used here was distilled and the pH is about 6.6.

Methods:

Fluorescence

Fluorescence spectroscopy, with pyrene as photosensitive probe, has been used to determine the polymer conformation in solution by measuring the polarity parameter (I_3/I_1) and coiling index (Ie/Im)[11-13]. It has been well known that as the pyrene molecule experiences a change in its environmental hydrophobicity, the relative intensities of the third (I_3) and first (I_1) vibrational bands of the monomer emission are affected. The ratio, I_3/I_1 (also known as polarity parameter), is a measure of the effective polarity of the medium where the pyrene molecule is located. A low value of I_3/I_1 indicates a hydrophilic environment and a high value hydrophobic environment respectively. Thus when the polymer molecules are streched the pyrene I_3/I_1 ratio should be low; when the molecules are coiled and/or form inter-molecular aggregates the ratio is high. The coiling index (Ie/Im), on the other hand, is an indication of the coiling level of polymer in solution. When pyrene molecules have a chance to meet each other, an excited state pyrene Py* can interact with a ground pyrene Py to form an excimer Py_2*. The ratio of pyrene excimer to monomer emission intensities (Ie/Im) is a useful measure of excimer concentration, and hence of local pyrene concentration. It is clear that a coiled polymer molecule will provide more opportunities for pyrene molecules to see each other than a stretched molecule would. Therefore, a coiled molecule will give a higher Ie/Im value than a stretched one. That is why the Ie/Im has been called "Coiling Index." If polymer molecules form aggregates, the Ie/Im value will go up based upon this principle.

All the fluorescence works in this report were carried out on a Photon International PTI-LS 100 spectrometer. A pyrene stock solution was prepared by stirring solid pyrene in water for 24 hours and filtering off the excess probe. This solution was added as required to the polymer solutions, for which the pyrene concentration was kept constant (11,12).

Viscosity

The viscosity was measured on TA Rheolyst AR1000-N at the shear rate of 1 s^{-1} at 25°C.

Light Scattering

Dynamic Light scattering (DLS), also known as photon correlation spectroscopy (PCS) and quasi-elastic light scattering (QELS), was run using a Brookhaven research-grade system with a BI-9000AT corrector and BI-200SM goniometer with adjustable angles of detection from 15° to 155°. A water-cooled Lexel Argon laser light source was used at a wavelength of 488 Angstrom. The samples were temperature controlled to +/- 0.1 °C and a refractive index matching liquid was used to reduce light bending at the glass interfaces. The measurements were done at a constant 90° angle in this work (14,15) and the particle size was recorded as mean hydrodynamic diameter.

Results and Discussions

Figure 2 shows the pH effect on the viscosity of the 106 HE polymer solution at a fixed concentration of 0.1 wt%. It is seen that the solution viscosity is extremely dependent on the pH. Below about pH 7.8, the viscosity is low and almost keeps constant. Above ~ pH7.8, the viscosity increases sharply and reaches a high plateau in a very small range of the pH growth. The plateau viscosity is about 7 times higher than the solution at low pH. This is believed due to cross-linking. Similar results can be found for most of these kinds of thickeners (1-5).

In figure 3, we plot the results of the natural pH of the solution as concentration is varied. As expected the solution has a high pH at diluted concentration and the pH decreases sharply with the increase of the concentration. Above about 0.1 wt% the solution is approaching the equilibrium reflected by the slight pH change. It is then possible to study the variation of the molecule conformation (or structure) with solution pH at different

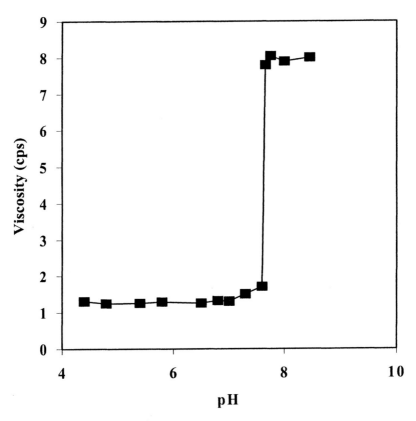

Figure 2. pH effect on the viscosity of the polymer solution at 0.1 wt% at 25 °C

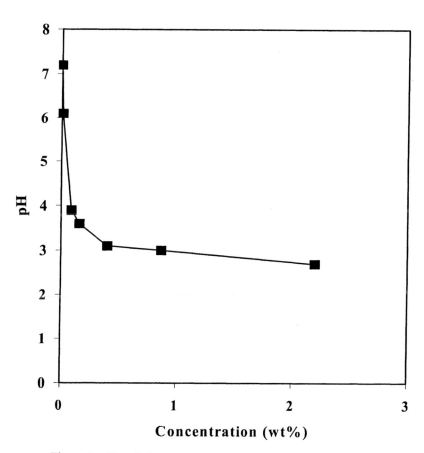

Figure 3. pH variation with the polymer concentration at 25 °C

concentrations, especially when the concentration is below 0.1 wt%. Therefore, without controlling the solution pH (at natural pH), we determined the particle size (mean hydrodynamic diameter) by using dynamic light scattering, pyrene polarity parameter (I_3/I_1) and coiling index (Ie/Im) on fluorescence at different concentrations at 25 °C.

As shown in figure 4, the particle size reduces strikingly as the concentration increases in the very diluted range, suggesting that the polymer molecules alter their structure from stretched to coiled with the pH decrease that is caused by the concentration increase. However, when the concentration is above about 0.02 wt% the particle size reduction is much slower with further addition of the polymer and eventually the mean hydrodynamic diameter approaches a constant value. This can be interpreted by considering that the coiling level of the molecules is restricted due to the molecular chemical configuration. Although the pH continued to decline with the concentration (far below 0.1 wt%), further alteration of the structure becomes difficult.

On the other hand, the inter-molecular reaction is also impossible because of the high coiling level of the molecules at that pH condition. This particle size result and the interpretation is confirmed by the pyrene fluorescence polarity parameter (I_3/I_1) measurement as can be seen in Figure 5. The I_3/I_1 ratio goes up quickly with the increase of the concentration in the diluted solution, indicating that the molecules change their conformation to the more coiled level. Again the limited coiling level leads to the constant I_3/I_1 ratio when the concentration is higher than about 0.02 wt%.

Figure 6 gives a direct indication of how the coiling level of the molecule varies with the concentration. The ratio Ie/Im increases with the concentration below 0.02 wt%, and approaches a constant value above 0.02 wt% consistent with the same mechanism as described above (the higher the Ie/Im value, the higher the coiling level of the molecules in solution). Both I_3/I_1 and Ie/Im results correspond to the particle size result. Above a concentration of 0.02wt% (or below ~ pH 5.6) the polymer molecules are entirely coiled and result in a minimum molecular size. With further increases the concentration the molecular structure as well as the size does not change any more.

To form a cross-link network in solution, the polymer molecules have to be in stretched conformation and the concentration has to be high enough to allow the molecules to see each other. In this case the critical pH is about 7.8 for 106 HE, above which the molecule is completely stretched and a cross-link can be formed (see figure 2). By controlling the solution pH at 8.2, we examined the aggregation properties of the polymer in solution. Since at pH 8.2 the molecules are completely stretched, the I_3/I_1 ratio is only attributed to the molecular aggregation. The continued increase of the pyrene I_3/I_1 ratio, as can be seen in figure 7 is, therefore, entirely due to the formation of aggregates. However, it is hard to distinguish whether the polymer forms more aggregates or the aggregates

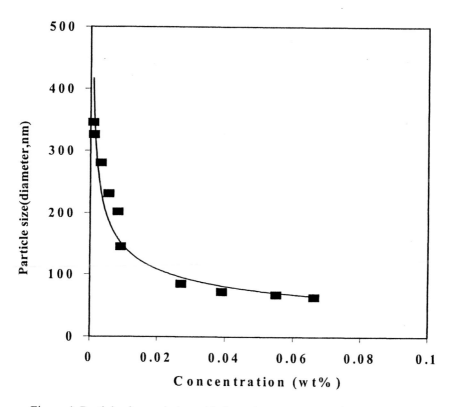

Figure 4. Particle size variation with the polymer concentration at natural pH

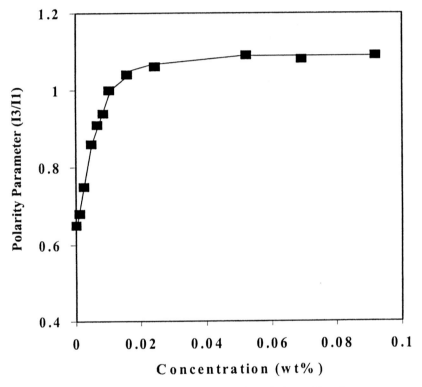

Figure 5. Pyrene polarity parameter (I3/I1) variation with polymer concentration at natural pH

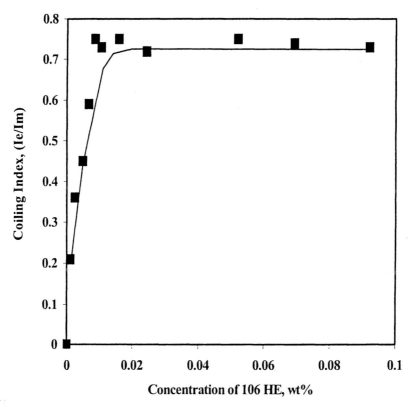

Figure 6. Coiling Index (Ie/Im) variation with polymer concentration at natural pH

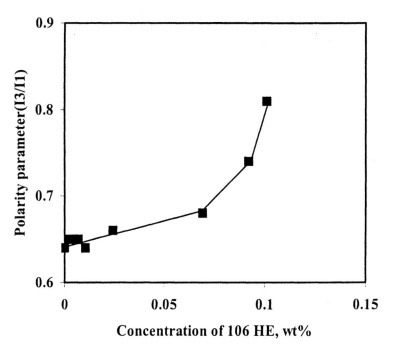

Figure 7. Pyrene polarity parameter (I3/I1) variation with polymer oncentration at pH 8.2

become bigger with the increase of the concentration by figure 7 alone. Therefore, we measured the particle size under the same condition, and the result is shown in figure 8. Remarkably, the particle size increases sharply at diluted range, but approaches a constant value with further increase of the concentration.

The mechanism of this phenomenon is considered as due to the aggregation restriction of the polymer in solution. In other words, there is a limitation for the size of aggregates to grow to. Beyond this limitation the molecules start to form new aggregates instead of enlarging the existing aggregates with the increase of the concentration before cross-linking. We have found that it is impossible to determine the particle size at even higher concentration because of the interaction among the aggregates at higher concentration, which eventually leads to cross-linking.

Figure 9 shows the pH effect on the pyrene I_3/I_1 ratio at a fixed concentration of 0.1wt%. The mechanism is similar to the concentration effect on the I_3/I_1 ratio as described above. At low pH the molecules are coiled, which results in a high I_3/I_1 ratio. With the increase of the pH, the molecules have gradually stretched out leading to a lower ratio. However, when the molecules stretch out, the interaction among them becomes possible at that relatively high concentration. Bearing in mind that this inter-molecular reaction is a positive contribution to the I_3/I_1 ratio. The compromise of the single molecule stretched out (reduce I_3/I_1 ratio) and the inter-molecule reaction (increase I_3/I_1 ratio) results in the apparent I_3/I_1 ratio decrease, albeit slowly.

Based upon the mechanism described above, we introduced an appropriate amount of the 106 HE into our water-based lithographic inks at a proper pH. The ink performance in sense of toning, misting and transferring has been significantly improved. It can be concluded that polymer conformation is a crucial criterion through which not only can we better understand the performance of suspension systems (including inks and paints, etc.), but also how to manipulate the formulae to the desired properties. Since polymers have been widely used in most printing inks and paints, it is believed that more attention will be paid to polymer conformation studies in the future, specifically their effects on network formation.

Summary

The conformation of the ter-polymer, 106 HE, varied from stretched to coiled with the decrease of the solution pH, or the increase of the concentration. This structure change correlates with the particle size alteration. At concentration 0.02 wt% (the corresponding pH 5.6) the polymer molecules reach the minimum coiled size and cannot be reduced further with an increase of the concentration

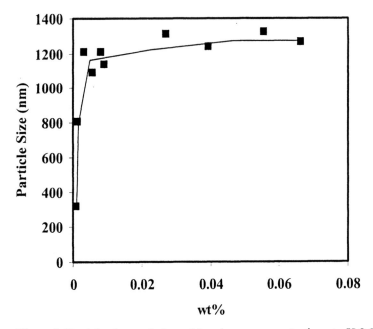

Figure 8. Particle size variation with polymer concentration at pH 8.2

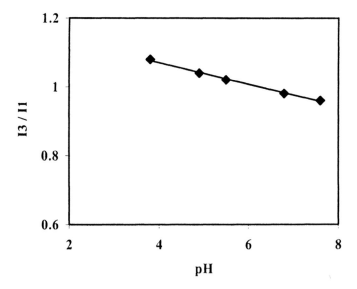

Figure 9. pH effect on the pyrene polarity parameter of the polymer solution at 0.1 wt%

or a decrease in pH. When the solution pH is equal to 7.8 or higher, the molecules are completely stretched out, and the aggregates eventually cross-links may form if the concentration is high enough. However, it was observed that the aggregation number of this polymer is restricted. Once the aggregate grows to a certain size, addition of more polymer leads to the formation of new aggregates instead of the expansion of old ones.

Acknowledgment

We wish to thank Langmuir Center for Colloids and Interfaces, Columbia University in the City of New York, where most of our work was conducted. We also wish to thank Professor Somasundaran, the director of the center, in supporting us to complete the work promptly.

References

1. Hulden, M.; *Colloids and Surfaces A*, 1994, 82, 263
2. Sperry, P.R.; *J. Colloid and Interface Sci.,* 1984, 99, 97
3. Shay, G.D.; *American Paint & Coatings J.,* 1993, 78(28), 44
4. Ning Wu, Ph.D. Thesis, Lehigh University, 1996
5. Olesen, K.R.; Shay, G.D.; and Rex, J.D.; *American Paint & Coatings J.,* 1995, Vol 79, No. 31, 51-59,
6. Shay, G.D.; Stalling, J.L. and Manus, P. J-M; *Surface coatings International,* 1997, 6, 285,
7. Glass, J.E.; Proceedings of the TAPPI Coating Conference, 1988, 287
8. Shay, G.D.; Bassett, D. and Rex, J.D.; JOCCA, *Surf. Coat. Int.,* 1993, 76 (11), 446-453
9. Bieleman, J.H.; Riesthuis, F.J.J. and van der Velden, P.M.; *Polym. Paint Colour J.,* 1986, 176, 450
10. Campbell, A.; Pan Z. and Somasundaran, P.; *Polymers in Particulate Systems;* Hackley, V. et al Ed., 2001, 135-155
11. Lou, A.; Pethica, B. P.; Somasundaran, P. and Fan, A.; *J. Dispersion Science and Technology,* 1999, 20, 9569
12. Qiu, Q.; Lou, A.; Somasundaran, P. and Pethica, B.A.; *Langumir,* 2002, 18, 5921

13. Sivadasan, K. and Somasundaran, P.; *Colloids and Surfaces,* 1990, 49, 229

14. Magid, L.; *Dynamic Light Scattering*; Clarendon Press, Oxford, 1993, 554-593,

15. Wines, T. H. and Somasundaran, P.; *J. Colloid and Interface Science*, 2002, 247

Chapter 15

Effects of Process Variables and Their Interactions on Rheology of Concentrated Suspensions: Results of A Statistical Design of Experiments

H. El-Shall[1], W. H. Kim[1], A. Zaman[2], S. El-Mofty[3], and I. Vakarelski[2]

[1]Department of Materials Science and Engineering and [2]Engineering Research Center for Particle Science and Technology, University of Florida, Gainesville, FL 32611
[3]Mining Engineering Department, Cairo University, Cairo, Egypt

Control of the rheological behavior and stability of concentrated dispersions is critical for the successful manufacturing of high quality products of particulate suspensions. Thus, a basic understanding of the role of different variables such as dispersant dosage, volume fraction of the solids, particle size distribution, and colloidal forces on flow properties of suspensions is required. In the past, studies have been conducted using one-variable-at-a time research strategy. In this research, statistical design of experiments used to provide predictive tools for the behavior of the system under various conditions. In this paper, role of dispersant dosage, solids loading, and pH of the suspension is studied.

Introduction

Advanced ceramics are used in many fields, such as electronic, magnetic, optical, mechanical, thermal, biological and aesthetic products (*1*). General requirements for these applications are high and uniform densities for green body and final sintered products (*2, 3*). To avoid excessive shrinkage during fluid removal and/or densification, molding methods require slurries containing the highest possible fraction of particles (*4*). It is well known that processing high solid content slurries poses several challenges including high values of rheological parameters such as yield strength, viscosity, etc. It is also equally known that such rheological properties could be modified by changing the dispersion/aggregation characteristics of the colloidal suspensions. In this regard, an enormous amount of research (*5-9*) has been done to disperse particles and overcome attractive van der Waals force in aqueous media environment by using electrostatic and steric repulsive force. This research also focuses on modification of these forces to achieve optimum processing conditions. In past studies, however, one-variable-at-a time research strategy has been used. Besides being time consuming, this research strategy does not yield interaction effects as well as optimization opportunities. On the other hand, statistical design of experiments overcomes these disadvantages and mathematical models can be developed to provide predictive tools for the behavior of the system under various conditions.

In this paper, role of dispersant dosage, solid loading, and ionic strength of the suspension is studied. The main and interaction effects of these variables on slurry viscosity and adsorption density of the studied dispersant are presented. Surface response methodology is employed to determine the conditions that may be used to produce highly concentrated slurries of minimal viscosity.

Experimental

Materials

Alumina A16SG (Alcoa chemicals, PA) of 99.8% purity, average BET surface area of 7.9m^2/g and mean particle size of 0.47 micron was used. Sodium salt of polyacrylic acid with a molecular weight equal to 2100 (Polysciences) was tested as dispersant. Sodium chloride (Fisher Scientific) was added to control ionic strength.

Slurry Preparation

Alumina powder was poured into a 150 ml glass flask containing a dispersant solution at required dosage and then was mixed thoroughly by shaking using a wrest-hand shaker for 1.0 hour. The suspension was sonicated using a probe type sonicator, for 75 seconds at 115W, and then sodium chloride was added to adjust the ionic strength, and finally the samples were agitated for 13 more hours.

Viscosity Measurements

Viscosity was measured using a modular compact rheometer with a concentric-cylinder measurement system (Models MCR 300 AND CC27, Paar Physica USA, Inc., Edison, NJ) having an inner cylinder diameter of 27 mm. Flow curves were determined at shear rates ranging between 0.01 and 1000/sec. It was important to pre-shear the samples at 600/sec for 40 seconds to remove air bubbles. The viscosity was then measured (30 points with 10 seconds duration between points).

Zeta Potential Measurements

Zeta potential was measured by using Accoustosizer (Matec Applied Science Inc., Hopkinton, MA). Since the natural pH values of the suspensions were mostly in the alkaline range, only HCl was used to lower the pH to the desired values.

Adsorption Density Measurements

After centrifuging alumina suspensions, the supernatant was used for determining residual polymer concentration utilizing Tekmar-Dorhman Phoenix 8000 TOC (Total Organic Carbon) Analyzer.

FTIR Spectroscopy

Sediments from centrifuged samples were dried at room temperature for 24 hours, then further dried at 50°C for 1hour. Infrared measurements were conducted using TNicolet MAGNA 760 Bench with Spectra Tech Continuum IR Microscope. To obtain absorbance peak of Na-PAA, a solution of Na-PAA in D.I. water was used in a silicon liquid cell.

AFM Measurements

Digital Instruments Nanscope III in a fused silica liquid cell was used in AFM measurement. Forces normal to the flat surface were measured according to the method described by Drucker et al. (*10*). Alumina plate and Silicon Nitride tip were used in this measurement. Before the force measurement, 30 minutes of adsorption time were used in this study.

Statistical Design

The statistical design used for viscosity and adsorption density measurements is given in Table I. The studied variables included solids loading and dispersant's (PAA) dosage. The pH was kept constant at 9.65±0.15, which was the natural pH. After measuring viscosity and adsorption density, the data were analyzed by Design-Expert 6.0.5 (*11*).   Viscosity values at 92.8 sec^{-1} shear rate are used for data analysis.

Result and Discussion

Zeta Potential Measurements and Calculated Force-Distance Curve

Zeta potential values of alumina suspensions in presence and absence of different dosages of the dispersant sodium salt of PAA, are given in Figure 1 as a function of pH. In these experiments, 0.03M NaCl was added. The data show that the isoelectric point of alumina is at about pH of 8.5 in absence of dispersant. This value agrees with that obtained by other investigators. It is important to note, however, that when PAA is added, the isoelectric point has shifted towards the acidic pH range as the dosage is increased. This behavior is also in agreement with other findings (*6, 12, 13*). The change in isoelectric point may indicate a specific adsorption of PAA on alumina surface as discussed later in this paper. Considering the zeta potential values at different pH ranges, we may be able to predict the dispersion behavior of these suspensions. For example, in the acidic pH range (around pH 4.0) the particles are highly positively charged. Thus, it is expected that the electrostatic repulsive forces will be larger than van der Waal forces leading to electro-statically dispersed slurry. This is clear from the calculated total forces shown in Figure 2. Based on the shown barrier, it is expected that the suspensions will be dispersed and the viscosity of high solid content slurries could be measured as shown in Figure 3.

On the other hand, at pH values close to the isoelectric point, van der Waal forces are expected to overcome the weak electrostatic forces as can be seen in Figure 4. Thus, slurries prepared in these pH ranges are expected to have very high viscosity values and high shear yield strength. Several researchers (*14, 15*) have experimentally proven this finding. Thus, it is expected that processing

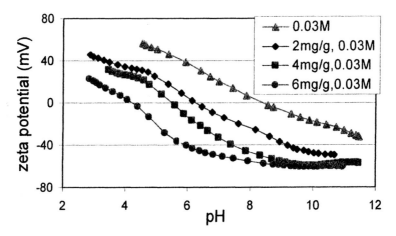

Figure 1. Zeta potential of 3.0Vol.% alumina suspensions at different dispersant dosages and in presence of 0.03 M NaCl

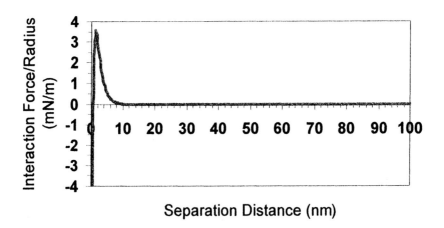

Separation Distance (nm)

Figure 2. Calculated sphere/sphere interaction curve at pH=4.0, zeta potential=58mV & 0.03M of added NaCl

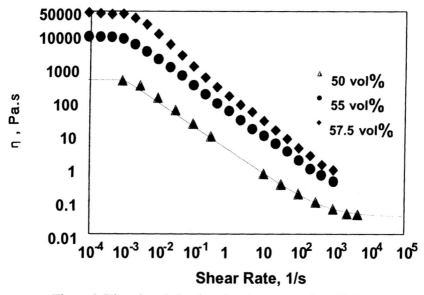

Figure 3. Viscosity of alumina slurries prepared at pH 4.3.
(Reproduced with permission from Chem. Engr. Commun. 1998, 169 (Oct./Nov.) 203–221. Copyright 1998 Taylor and Francis.)

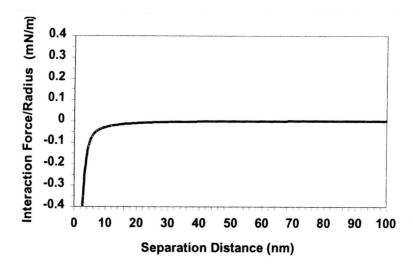

Figure 4. Calculated sphere/sphere interaction curve at pH=9.8, zeta potential=-16mV & 0.03M of added NaCl

slurries at these pH values may require use of low solids loading or the use of dispersants. Such dispersants may induce repulsive forces (steric and/ or electrosteric) depending on dispersant structure, molecular weight, conformation in the bulk and at the surface, etc.

The increase in the negative zeta potential values of alumina suspensions upon addition of PAA in the alkaline pH range is obvious from Figure 1. Such increase could be attributed to adsorption of negatively charged fully ionized PAA molecules at this pH range (pK_a of PAA is around pH 4.5).

Table I. Statistical design with central point for viscosity measurements

Run #	Solid loading	Dispersant dosage
1	—	—
2	+	—
3	—	+
4	+	+
5	—	o
6	+	o
7	o	—
8	o	+
9	o	o
10	o	o
11	o	o

Variables Levels:	Low (-)	Middle (0)	High (+)
Solids Loading (Vol%)	40	45	50
Dispersant dosage (mg/g)	2	4	6

PAA Adsorption on Alumina at High pH Range

PAA adsorption isotherms on alumina in solutions of low and high ionic strength values are given in Figure 5. The data indicate that ionic strength does not have a significant effect on adsorption density. Also, this type of isotherm is similar to the polyelectrolyte adsorption isotherms as discussed by other researchers (5, 6, 16). Since the PAA is highly negative in this pH range and the alumina surface is also negatively charged, then such adsorption could be due to complexation with Al^{+3} ions on alumina surface as suggested in literature (17). To confirm such adsorption mechanism, FTIR spectra of PAA, in presence and absence of alumina particles, was obtained. Figure 6 shows FTIR spectra of Na-PAA at pH=8.5. The sharp peak at 1557 cm^{-1} is due to the asymmetrical stretching of COO-. Figure 7 depicts the FTIR spectra of pure alumina and alumina coated with PAA at dosages of 2mg/g and 6mg/g at 0.01M of added NaCl. It can be seen that after adsorbing PAA on alumina, the absorbance peak 1557 cm^{-1} of asymmetrical stretching of COO- in Na-PAA is shifted to 1576 cm^{-1} (18). This may explain the polymer adsorption on negatively charged alumina surface under these research conditions.

Adsorption of PAA has resulted in an increase in negative charge as well as increase in repulsive forces between AFM tip and Alumina plate as can be seen in Figures 1 and 8. It is important to note that in absence of PAA, the forces are attractive as expected from the low zeta potential values at this pH range. Most importantly, the adsorbed polymer does not desorb by washing as indicated by the strong repulsive forces even after washing alumina surface by NaCl solution. This confirms the complexation mechanism as mentioned above and described in literature (17).

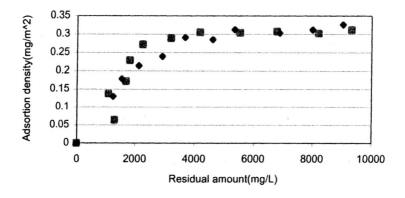

■ 0.01M NaCl   ◆   0.2518M NaCl

Figure 5. Adsorption isotherm in 40Vol% alumina suspensions at pH=9.70±0.10

Statistical Design Analysis of Viscosity Measurements

Rheological Response

Viscosity of alumina slurries is measured at different shear rates. Data for low and high ionic strength conditions are given in Figure 9. It is interesting to note that all slurries show a shear thinning behavior. In other words, any network structure built in these slurries breaks down as shear rate is increased. In addition, viscosity of up to 50 vol.% slurries (at a pH value close to the isoelectric point) could be measured. This is attributed to the electrosteric repulsion induced by PAA adsorbed molecules (Figure 8). The viscosity values at high solid content is higher than at lower solid content slurries due to increased probability of network and structure formations in these slurries as indicated by other researchers (*9, 19*). However, the viscosity values obtained at high ionic strength are much higher than that at low ionic strength solutions.

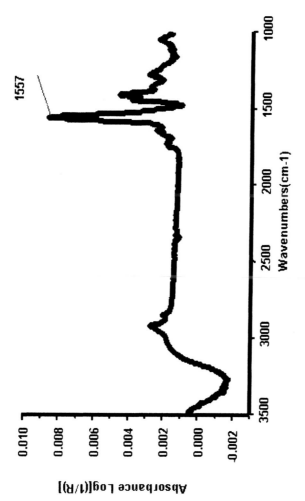

Figure 6. FTIR spectra of Na-PAA at pH=8.5

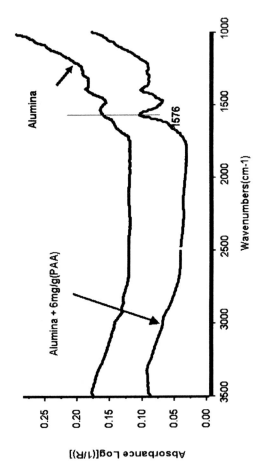

Figure 7. FTIR spectra of alumina with and without added Na-PAA

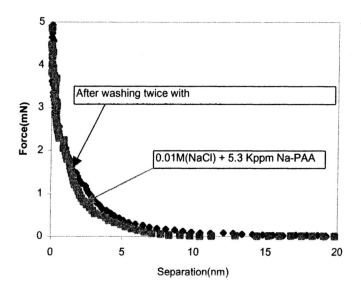

Figure 8. Atomic force measurement curves at pH=9.80

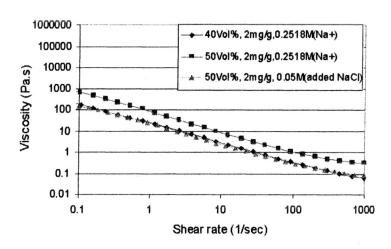

Figure 9. Viscosity of alumina slurry at different conditions at pH=9.65±0.15

This could be due to changes in polymer conformation due to high content of sodium ions. It should be remembered, however, that ionic strength did not significantly affect amount of PAA adsorbed on alumina as mentioned above (Figure 5). Increasing PAA dosage has resulted in decrease in viscosity as can be seen from Figure 10. However, further increase in PAA dosage beyond 3.5 mg/gm, did not produce a significant decrease in viscosity. This is not surprising since the saturation adsorption value is reached at this dosage as can be seen in Figure 5.

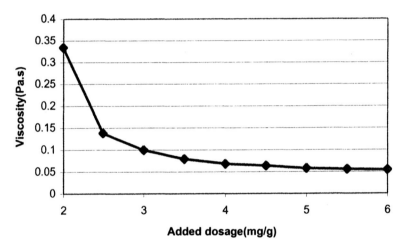

Figure 10. Viscosity measurement at different added Na-PAA dosage Alumina A16SG(40Vol%), 0.2518M(Na⁺), pH=9.60±0.1

Viscosity values at 92.4/sec shear rate were used for statistical analysis at each statistical design condition. Figure 11 is the contour plot of viscosity values that could be obtained as the solids loading and PAA dosage are changed. The data clearly show that viscosity increases as solids loading is increased especially at low levels of PAA additions. Increasing PAA minimizes effect of solids loading due to the dispersion effect of PAA. It is important to notice here also that after 3.5mg/ gm viscosity reaches a low value that is represented by a wide valley at the level of 73.3 mPa.s. This again confirms the previous finding that higher dosages of PAA are not required either for adsorption or viscosity reduction.

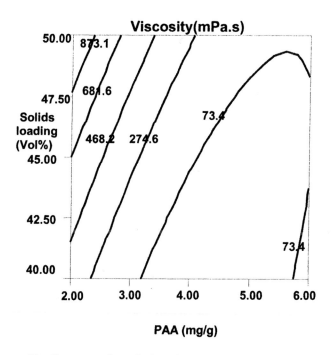

Figure 11. Contour plot of viscosity of alumina slurries(pH 9.6+/-0.1) as a function of Na-PAA dosage and solids loading at 92.4/sec shear rate (R^2=0.9608, experimental error=1.76)

Summary and Conclusions

Alumnia slurries of high solid contents could be prepared in a stable condition at low pH values where electrostatic repulsive forces induce stabilization of these slurries. Closer to the isoelectric point (pH of 8.6) slurries are coagulated and of high viscosity that may render processing of such slurries difficult. Thus, dispersants should be added to obtain high low viscosity-solids content slurries. In this regard, PAA is a good dispersant for such slurries. Due to its specific adsorption on alumina surface, it increases the repulsive forces by the electrosteric mechanism.

It is also important to note that ionic strength was found to affect rheological behavior of alumina slurries. Ionic strength is proposed to affect the conformation of the polymer in bulk of solution leading to increased viscosity at high ionic strength values. FTIR, AFM, Zeta potential measurement, and adsorption isotherms are used to confirm the above conclusions. Statistical design of experiments is used to generate surface response of viscosity as a function of solids loading and PAA dosage. The response surface generated from this design is proved helpful in identifying the conditions that may be used to achieve low viscosity and high solid content slurries.

Acknowledgment

We acknowledge the financial support from Department of Materials Science and Engineering and the Engineering Research Center for Particle Science & Technology (ERC) at the University of Florida. Gratitude is due to Prof. Somasundaran of Columbia University, our colleagues Drs. B. M. Moudgil, W. Sigmund, Y. Rabinovich, and graduate students: P. Singh, S. Brown and M. Esayanur, for their helpful discussions.

References

1. Reed, J.S., "Principles of Ceramics Processing," *Wiley Interscience* 1994.
2. Sacks, M.D., Lee, H.W., Rojas, O.E., "Suspension Processing of Al_2O_3/SiC Whisker Composites," *J. Am. Ceram. Soc.* 71[5] (1988) 370-379.
3. Sacks, M.D., Lee, H.W., Rojas, O.E., "Pressureless Sintering of SiC

Whisker-Reinforced Composites," *Ceram. Eng. Sci. Proc.* 9[7-8] (1988) 741-754.

4. Lange, F.F., "Powder Processing Science and Technology for Increased Reliability," *J. Am. Ceram. Soc.* 72[1] (1989) 3-15.

5. Cesarano, J., Aksay, I.A., "Stability of Aqueous α-Al$_2$O$_3$ Suspensions with Poly (methacrylic acid) Polyelectrlyte," *J. Am. Ceram. Soc.* 71[4] (1988) 250-255.

6. Hackley, V.A., "Colloidal Processing of Silicon Nitride with Poly (acrylic acid) : I, Adsorption and Electrostatic Interactions," *J. Am. Ceram. Soc.* 80[9] (1997) 2315-2325.

7. Hackley, V.A., "Colloidal Processing of Silicon Nitride with Poly (acrylic acid): II, Rheological Properties," *J. Am. Ceram. Soc.* 81[9] (1998) 2421-2428.

8. Cho, J.M., Dogan, F., "Collidal Processing of Lead Lanthanum Zirconia Titanate Ceramics." *J. Mat. Sci.* 36 (2001) 2397-2403.

9. Zaman, A.A., Moudgil, B.M., Fricke A.L., El-Shall, H., "Rheological Behavior of Highly Concentrated Aqueous Silica Suspensions in the Presence of Sodium Nitrate and Polyethylene Oxide," *J. Rheol.* 40[6] (1996) 1191-1210.

10 Ducker, W.A., Senden, T.J., Pashley, R.M., *Nature*, 353 (1991) 239.

11. Stat-Ease, Inc. Hennespin Square 2021 E.Hennespin Ave, Minneapolis, MN, 55413.

12. Leong, Y.K., Scales, P.J., Healy, T.W., Boger, D.V., "Interparticle Forces Arising From Adsorbed Polyelectrolytes in Colloidal Suspensions," *Colloids Surf. A: Physicochem. Eng. Aspects* 95 (1995) 43-52.

13. Tjipangandjara K.F., Somasundaran, P., "Effects of Changes in Adsorbed Polyacrylic Acid Conformation on Alumina Flocculation," *Colloids Surf.* 55 (1991) 245-255.

14. Scales, P.J., Johnson, S.B., Healy, T.W., Kampur, P.C., "Shear Yield Stress of Partially Flocculated Colloidal Suspensions," *AIChE J.* 44 (1998) 538-544.

15. Leong, Y.K., Boger, D.V., Scales, P.J., Healy, T.W. Buscall, R., "Control of the Rheology of Concentrated Aqueous Colloidal Systems by Steric and Hydrophobic Forces, " *J. Chem. Soc.*, Chem. Commun. 7(1993) 639-641.

16. Davies, J., Binner, J.G.P., "The Role of Ammonium Polyacrylate in Dispersing Concentrated Alumina Suspension," *J. Euro. Ceram. Soc.* 20 (2000) 1539-1533.

17. Bjelopavlicm M., El-Shall, H., Moudgil, B.M., "Role of Polymer Functionality in Specific Adsorption to Oxide: A Molecular Recognition Approach," *Polymers in Particulate Systems Surfactant* science series volume 104, Marcel Dekker.

18. Liu, Y.O. and Guo, J.K., "Adsorption of Acrylic Copolymers at the Alumina-water Interface," Colloids Surf. A: Physicochem. Eng. Aspects, 164 (2000) 143-154.
19. Jefferey, D.J. and Acrivos, A., "The Rheological Properties of Suspensions of Rigid Particles," AIChE J. 22 (1976) 417-432.

Chapter 16

Stability of Highly Loaded Alumina Slurries

S. Chander and M. Chang

Department of Energy and Geo-Environmental Engineering,
The Pennsylvania State University, University Park, PA 16802

Stability of highly loaded alumina slurries was studied as a function of reagent type and concentration. The reagents investigated include several ABA and AB type nonionic block copolymers. The stability profiles of various slurries were mapped by measuring the position of the 'sediment line' and the 'mud line' as a function of time. A new method to detect sediment and mud lines is described. Using this method the stability measurements were made as a function of both reagent concentration and molecular weight, within homologous series of reagents. A critical stability parameter was defined as the time at which the dispersed phase just disappears. This parameter was used as one of the stability parameters. The other parameter used to characterize the sediment after prolonged settling was the volume percent solids in the sediment. The effect of reagent concentration and molecular weight on these two parameters is discussed.

Highly loaded colloidal suspensions are important in many industries such as ceramics, coatings, paints, pesticides, water treatment, mineral processing, pharmaceutical and food industry. With increased emphasis on environmental concerns, the dispersion of particles in aqueous media is getting more emphasis. Low and high molecular weight polymers are frequently used as dispersants and wetting agents to modulate flow and stability properties. The stability of aqueous dispersions in the presence of various polymers has been extensively studied from a practical and a theoretical aspect by many researchers, the mechanism of the stabilization is still not well elucidated, and the practical solution of most dispersion problems is still empirical.

Ionic reagents have been widely used as dispersants or stabilizers to obtain electrostatic stabilization of aqueous alumina suspensions (*1*), while the use of nonionic reagents for aqueous dispersion is less common (*2*). The possible reason is that nonionic reagents alone do not improve stability and rheological properties of aqueous alumina suspensions; therefore, a combination of electrostatic and steric stabilization is required for optimal performance.

In this study, citric acid was selected as the reagent to obtain electrostatic stabilization. It is known to be a good dispersant for alumina. The amount needed to produce low viscosity alumina suspension is very small. Also, it does not introduce undesirable species such as nitrogen, sulfur, etc. It is also biodegradable if discharged into the environment as process fluid. The effect of molecular stucture of nonionic surfactants on stability of electrostatically stabilized suspensions was investigated and the results are discussed.

Materials and Methods

High purity alumina powder (AKP-50, Sumitomo Chemical Co., Japan) was used in this study. Density, BET surface area, and particle median size of this alumina powder were reported by the manufacturer as 3.99 g/cm^3, 10.1 m^2/g, and 0.3 μm respectively. The PZC (point of zero charge) of this alumina was found to be pH 9.1.

The nonionic ABA-type triblock copolymers were received from BASF Corp. A list of the reagents used in this study is given in Table I. The ABA block copolymers were water-soluble and consist of a polypropyleneoxide block linked to polyethyleneoxide block at both ends; their structure is shown in Figure 1. Citric acid [$HO-C(COOH)(CH_2COOH)_2]H_2O$, MW 210.1] was obtained from J.T. Baker Chemical Co. The nonionic AB-type block copolymers used in this study were: polyethylene glycol-co-polypropylene glycol [$H(OCH_2CH_2)_x$ $(OCH_2CH(CH_3))_y$-OH] of MW 2500] containing 75% wt.% ethylene glycol and polyethylene glycols [$H(OCH_2CH_2)_n$-OH] of MW 1500 & 4000. Analytical grade HNO_3 and NaOH were used for pH adjustment. Distilled water was used throughout the invsestigation.

Alumina slurries of 75 wt. % (43 vol. %) were prepared by mixing 30 g alumina with 10 ml distilled water in 60 ml plastic bottles. The bottles were vigorously shaken by hand after the desired solution was added to adjust the pH. The slurries were further dispersed with a high intensity ultrasonic processor (Sonics & Materials Inc.) for 20 seconds. The desired amounts of reagents, based on the dry weight of alumina powders, were added and the bottles were mixed with a wrist action shaker (Burkell Corporation) for at least 5 hours before the measurements.

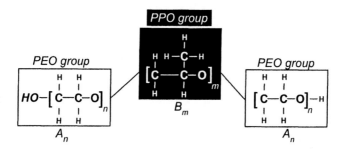

Figure 1. A schematic representation of ABA type surfactants. These surfactants are commercially available as Pluronic® series of reagents.

Table I. Selected properties of the surfactants used in this study. The data for ABA triblock copolymers was obtained from Vaughn and Dekker (3).

Copolymer	Formula	Molecular Weight	N_{PO}/N_{EO}	HLB
L-44	$E_{10}P_{21}E_{10}$	2200	1.05	12-18
L-64	$E_{13}P_{30}E_{13}$	2900	1.15	12-18
P-84	$E_{19}P_{39}E_{19}$	4200	1.03	12-18
P-104	$E_{27}P_{56}E_{27}$	5900	1.04	12-18
P-103	$E_{17}P_{56}E_{17}$	4950	1.65	7-12
P-65	$E_{19}P_{30}E_{19}$	3400	0.79	12-18
P-85	$E_{26}P_{39}E_{26}$	4600	0.75	12-18
P-105	$E_{37}P_{56}E_{37}$	6500	0.76	12-18
F-38	$E_{42}P_{16}E_{42}$	4650	0.19	>24
F-68	$E_{76}P_{30}E_{76}$	8400	0.20	>24
F-88	$E_{103}P_{39}E_{103}$	11400	0.21	>24
PAA(D-3031)	$[-CH_2CH(COOH)-)_x$	3000	-	-
NP-15	$C_9H_{19}-F-(OCH_2CH_2)_{15}.OH$	880	-	-
PEG/PPG	$H(OCH_2CH_2)_x(OCH_2CH(CH_3))_y-OH$	2500	-	-
PEG	$H(OCH_2CH_2)_x-OH$	1500, 4000		

Centrifugal Settling Test and Stability Profile

Centrifugal settling tests were performed by using a constant temperature centrifuge (Model: Centra MP4R, of International Equipment Company). In a typical experiment, suspensions were loaded into the 6.5 ml centrifuge tubes, and sealed by a wax paper film. The centrifuge was set to run at the speed from 2000 to 3500 rpm (500 to 1500 g) and various periods of time ranging from 10, 20, 30, to 60 minutes. After each centrifugation, the sediment height of the suspension was measured and the final sediment height was obtained to estimate solids content in the sediment. For comparison with gravitational settling, the centrifugation time was converted to equivalent gravitational settling time by the formula: $T = \omega^2 Rt/g$, where, T is the equivalent gravitational settling time (in days), ω is the centrifuge angular speed (in s^{-1}), R is the distance between the centrifuge rotor and the center of the suspension (in meter), and t is the centrifugation time (in days) g is the gravitational acceleration (in m/s^2). The time taken to accelerate and stop the centrifuge was usually less than 1 minute and was therefore neglected.

A simple method was developed in this study, as illustrated in Figure 2 to monitor settling behavior. After centrifugation, the sample tube was placed under a fiber optics light. Several zones could be observed due to the light absorption, scattering and diffraction. At the bottom was the sediment zone with a slightly bright white color. The next was the dispersed zone, which is considered to have the same solids content as the suspension before settling. Wedlock et al. (*4*) used an ultrasonic technique to confirm that the solids content in the dispersed zone remains essentially constant. The appearance of the dispersed zone was darkest because of the relatively small amount of scattered light. The turbid zone was composed of very fine particles that settled very slowly. The height of each zone was recorded as a function of time, t and the results from such measurements at several times were plotted to obtain the settling profiles as presented in Figure 2. It can be seen that the height of sediment zone increased and that of the dispersed zone decreased with time. For suspensions with high solids loading (~43 vol.%) the turbid zone was very thin (less than 3%) in most cases. The percent solids in this zone are negligible compared to those in the sediment and dispersed zone. As far as the stability is concerned, the dispersed zone is considered the most important because this zone is expected to have about the same solids content as the original suspension. Therefore, the time for disappearance of the dispersed zone marked as t* in Figure 2, was used as a measure of stability. The second settling parameter, that is volume percent solids in the sediment was determined from the volume of the sediment layer after completion of the settling test.

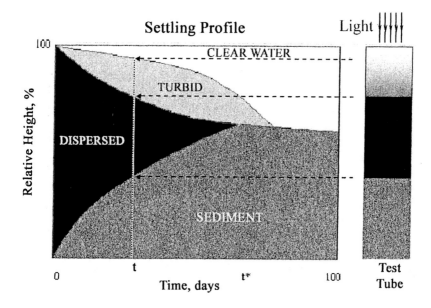

Figure 2. A schematic of the settling profile of suspensions obtained by centrifugal settling method. Various zone in the settling tube at time 't' are shown on the right and the settling profile is shown on the left.

Results and Discussion

Effect of pH. Suspensions containing 43 vol.% of alumina were dispersed by HNO_3 at pH 4, 5, and 6 were and the parameter t^* for the suspension was determined to be 125 days at pH 4. The final solids content in sediment were 63.6 vol.%. For the suspensions at pH 5 and 6, t^* could not be obtained because at these pH values, the alumina particles appear to be flocculated and the dispersed zone could not be determined. The final solids content in sediment for pH 5 and 6 were 61.6 vol.% and 56.4 vol.%. At pH 4, the electrostatic repulsion forces between alumina particles are sufficiently strong to disperse the particle. In separate measurements, viscosity and yield stress were found to be significantly lower at pH 4 compared to pH's of 5 and 6. The sediment packing after settling was also the highest compared to the suspensions at pH 5 and 6. The suspension at pH 6 was most flocculated, it had the highest viscosity and yield stress as well as the lowest sediment packing density.

Effect of Citric Acid Concentration. The value of the parameter t* obtained for suspensions in the presence of 0.2, 0.3 and 0.5% citric acid at pH 7 were 140, 200, 200 days respectively. The corresponding solids content for their sediments were 62.4, 60.1, and 61.5 vol.% respectively. The stability increased when citric acid concentration was increased from 0.2 to 0.3%. However, further increase in the citric acid concentration to 0.5% did not increase the stability. These results correlate well with the rheology properties. Both the viscosity and yield stress were higher for the suspension with 0.2% citric acid when compared to suspensions with citric acid concentration of 0.3% and 0.5%. These results indicate that a minimum of 0.3% citric acid is needed for good dispersion of alumina at pH 7.

Effect of Nonionic PEO/PPO/PEO Triblock Copolymers. The centrifugal settling tests were performed in the presence of various amounts of PEO/PPO/PEO triblock copolymers and the results are presented in Figures 3 and 4. The results show that the stability of suspension increased with increase in the polymer concentration and molecular weight of the reagent. No such trend was observed for % solids in the sediment. For each reagent a complex relationship was observed between % solids in the sediment and the reagent concentration. Two peaks were observed in the sediment packing vs. concentration plots shown in Fiures 3 and 4, especially for reagents containing 40% EO. The first maximum occurred at low concentrations (~0.5%) and the second at higher concentration of about 5%. Results with other reagents, shown in Figures 3 and 4 were similar but the effect of reagent type and concentration was more complex. At low reagent concentration, the alumina surface covered with the adsorbed polymer could lead to better packing. The EO groups are expected to adsorb at the alumina surface by hydrogen bonding leaving the PO groups extended in to the solution. When the suspensions settled, the hydrophobic PO groups around the alumina particles functioned as lubricant and thus generated denser sediment. A further increase in polymer concentration would result in multi-layer formation reducing percent solids in the sediment. On the basis if these results we consider that hydrophobic lubrication promotes sediment packing. The results in Figure 4 show that highest packing density is obtained with P104 that contains 40% EO. In comparison, regents with 30% and 50% EO groups gave lower sediment density.

The settling test results for the suspensions in the presence of various amounts of polymers containing 80% EO groups are also shown in Figure 3. The stability parameter increased with increase in the polymer concentration and molecular weight. For concentrations greater than 2%, the sediment packing was greater for lower molecular weight reagents. This trend is inverse of the one observed for reagents with 40% EO where higher sediment packing was observed for reagents with higher molecular weight.

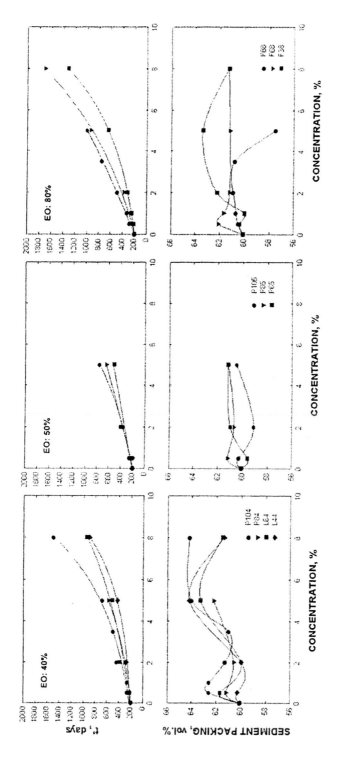

Figure 3. Settling test results of alumina suspensions in the presence of ABA triblock copolymers containing 40%, 50% and 80% EO groups.

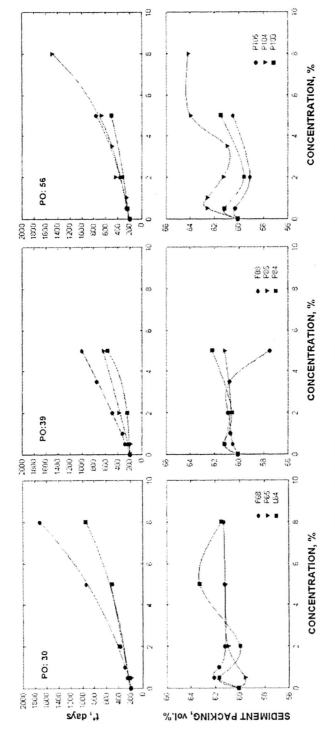

Figure 4. Settling test results of alumina suspensions in the presence of
ABA tri block copolymers containing 30, 39 and 56 PO groups.

Effect of Nonionic Diblock Copolymers. Two reagents, namely PEG/PPD diblock copolymer of molecular weight 2500 and Tergitol NP-15 of molecular weith 880 were used in this series of tests. These reagents did not have a very large effect on the stability but the effect on percent solidents in the sediment was significant as can bee seen from Table II. The suspension stability increased slightly with increase in the reagent concentration. The percent solids in the sediment increased at a reagent concentration of 0.5% with PEG/PPG and at 5% with Tergitol NP-15. At higher reagent concentrations, the suspensions are most likely flocculated resulting in a lower sediment packing density.

Table II. Effect of AB Type Nonionic Surfactants on Stability of Alumina (43% by volume).

Surfactant	Concentration	t*	Volume Percent Solids in Sediment
PEG/PPG	0.0	200	60.1
	0.5	200	62.3
	2.0	315	60.7
	5.0	395	60.8
Tergitopl NP-15	0.0	200	60.1
	0.5	200	60.3
	2.0	255	61.2
	5.0	420	62.9
	8.0	895	62.1

Effect of Polyethylene Glycol. The stability parameter, t* of suspensions in the presence of 5% PEG-1500 (MW: 1500) and PEG-4000 (MW: 4000) was 400 and 760 days, respectively. The solids contents in sediment after settling were 60.8 and 61.9 vol.%. The effect was considered to be small.

Summary Discussion

A centrifugal settling method was developed to map the stability profile of highly loaded slurries. Based on these measurements two parameters were used to define stability. These were: a critical settling parameter (t*) and volume percent solids in the sediment. The critical settling parameter, was defined as the time for disappearance of the dispersed layer.

The nonionic copolymers tested in this investigation increased the stability of highly loaded alumina slurries, as measured by the stability parameter, t*. This behavior was observed for both the slurries dispersed with nitric acid at pH 4 and citric acid at pH 7. To a large extent the increase was a function of the molecular weight and the reagent concentration as can be seen in Figure 5. The molecular structure of the reagent playing less important role on the settling parameter, t*. The data presented in Figure 6 shows that there is no significant correlation between percent solids in the sediment and molecular weight. The effect of reagent concentration on packing density was complex as dicussed in Figures 3 and 4.

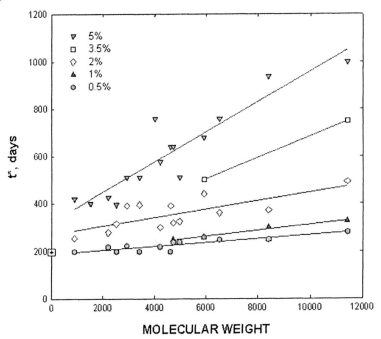

Figure 5. Effect of molecular weightof nonionic reagents tested in this study on the settling parameter. Varios symbols correspond to different reagent concentrations.

An increase in the stability with increasing concentration could be attributed to hydrophobic interaction forces between the particles covered with adsorbed polymers leading to steric interaction. As reported by Napper (4), the EO group is hydrophilic and can form hydrogen bond with carboxylic groups. Hidber et al. (5) reported that only one or two carboxylic groups of each citrate molecule attached to the alumina surface to form covalent bonds with alumina surface. Thus, the free carboxylic groups formed hydrogen bonds with the EO groups of

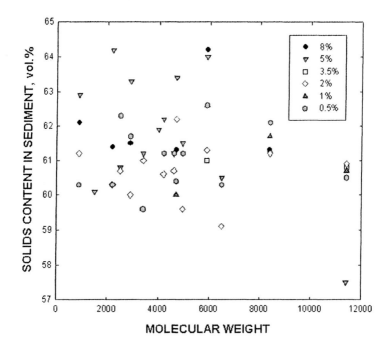

Figure 6. Correlation between percent solids in the sediment and molecular weight of nonionic reagents tested in this study. Varios symbols correspond to different reagent concentrations

the polymer. The PO groups of the polymer extended in to the solution. Because the PO groups are hydrophobic, they would interact with the PO groups of other particles or the free polymers in the solution. With an increase in reagent concentration the hydrophobic interactions between the PO groups increased resulting in higher stability. The interaction forces between the PO groups would be maximum when the particles are covered with a monolayer of adsorbed polymer. At higher concentrations, the PO groups interacted with the PO groups of the free polymers in the solution to form bilayer promoting stability.

The hydrophobic interactions between PO groups also increased lubrication between particle promoting increase in percent solids in the sediment. Highest packing density was obtained for ABA block copolymers with 86 PO groups and 40% EO groups.

250

Acknolegements

This research was supported by a grant from the Particulate Materials Center, The Pennsylvania State University, University Park, PA.

References

1. Cesarano, III J.; Aksay I. A., "Processing of Highly Concentrated Aqueous α-Alumina Suspensions Stabilized with Polyelectrolytes", *J. Am. Ceram. Soc.,* 71, **1988**, pp.1062-1067.
2. O. Omatete, and Alan Bleier, "Evaluation of Dispersants for Gelcasting of Alumina"" *Dispersion and Aggregation, Fundamentals and Applications*, B. M. Moudgil and P. Somasundaran, editors, Engineering Foundation, New York, **1994**, pp.269-278.
3. Vaughn, M. et. al., 1996, "Nonionic surfactants : polyoxyalkylene block copolymers", M. Dekker, New York
4. Napper, D.H. "Polymeric Stabilization of Colloidal Dispersions", Academic Press, London, New York, **1983**.
5. Hidber, P.C.; Graule, T.J.; Gauckler, L.J., "Citric Acid - A Dispersant for Aqueous Alumina Suspensions", *J. Am. Ceram. Soc.,* **1996**, 79 (7), pp. 1857-1867.

Indexes

Author Index

Subject Index